Praise for *Keeping Watch*

"O'Malley has given us an engaging and witty book that is a thorough piece of scholarship, amply documented, meticulously assembled and thought-provoking."

— *The Los Angeles Times Book Review*

"There is a fascinating story to be told here. . . . It goes beyond the mechanics of timekeeping itself and to the heart of the evolution of the American mind."

— *Chicago Tribune*

"Well-researched and thorough"

— *The New York Times Book Review*

"Entertaining and informative . . . *Keeping Watch* is full of rich explanations and commentary about measuring the day."

— *Newsday*

". . . as a catalogue of timekeeping prejudices and disputes, it is entertaining, and *Keeping Watch* does illuminate some of the foibles of America's young republic."

— *The New York Times*

"O'Malley's history of American time is a lively and readable narrative. His observations and interpretations are often astute and sometimes provocative. *Keeping Watch* offers readers a fresh look at a timely topic."

— *The Christian Science Monitor*

PENGUIN BOOKS

KEEPING WATCH

Michael O'Malley holds a Ph.D. in American history from the University of California at Berkeley. He is an assistant professor of American history at Vassar College.

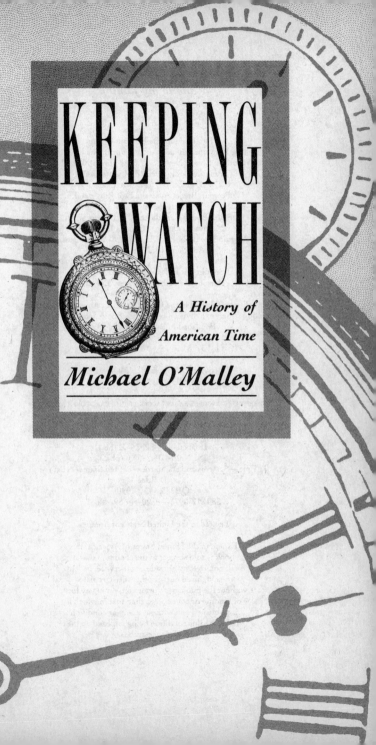

KEEPING WATCH

A History of

American Time

Michael O'Malley

To Kimberly,
and to my family

PENGUIN BOOKS
Published by the Penguin Group
Viking Penguin, a division of Penguin Books USA Inc.,
375 Hudson Street, New York, New York 10014, U.S.A.
Penguin Books Ltd, 27 Wrights Lane,
London W8 5TZ, England
Penguin Books Australia Ltd, Ringwood,
Victoria, Australia
Penguin Books Canada Ltd, 2801 John Street,
Markham, Ontario, Canada L3R 1B4
Penguin Books (N.Z.) Ltd, 182–190 Wairau Road,
Auckland 10, New Zealand

Penguin Books Ltd, Registered Offices:
Harmondsworth, Middlesex, England

First published in the United States of America by
Viking Penguin, a division of Penguin Books USA Inc., 1990
Published in Penguin Books 1991

1 3 5 7 9 10 8 6 4 2

THE LIBRARY OF CONGRESS HAS CATALOGUED THE HARDCOVER AS FOLLOWS:
O'Malley, Michael.
Keeping watch: a history of American time / Michael O'Malley.
p.   cm.
ISBN 0–670–82934–X (hc.)
ISBN 0 14 01.2433 0 (pbk.)
1. Time—Systems and standards—United States—History.
I. Title.
QB223.043   1990
389′.17′0973—dc20   89–40681

Printed in the United States of America

Preface

I wasted time, and now doth time waste me;
For now hath time made me his numb'ring clock.
　　　　　　　　　　　　—Richard II

Amerca, *Time* magazine announced in April 1989, is "running itself ragged." "These are the days of the time famine," the cover story insisted; everywhere you look you find Americans desperately trying to do more with less. Bosses demand longer and longer hours. Combining work and child rearing makes parents frantic, unable to keep up, and meanwhile the jet stream of new information rushes past, ever faster. Time-saving devices never seem to save us any time—computers, cellular phones, and fax machines make international business go round the clock. We live in a global community, *Time* pointed out, knit together by instantaneous communication and standard time. An architect complains that his staff managed to reach him while he sat on the beach at Anguilla, supposedly on vacation. "There are times," he says, "when our lives are clearly leading us." The "time famine" strikes both rich and poor—who has never glanced resentfully at the clock, racing a deadline, and not wondered how the machine became so powerful? According to *Time*, the average American now enjoys 37 percent less leisure than in 1973. Why this shortage of time?[1]

To explain our problem, we have to sort out the difference between what we think about work and how we understand *time*. Life in America has always been plagued by one peculiar fact: work nearly always expands to fill the time available. Time-saving devices and strategies for more efficiency only raise our expectations about what we can get done. Save time in one job, and some other task appears to fill that time up. Buy a microwave oven, and the time you gain heating up dinner goes to chores, or classes, or work brought home. We resent this tendency, but in fact most Americans believe strongly in work, and have a deep suspicion of loafers.

Eighty years ago Max Weber called this the "Protestant Ethic," the belief that idleness is sin and that work, pure and simple work, is virtuous for its own sake. Weber's classic analysis rooted modern society's obsession with saving time in the Calvinist sense of anxiety about salvation. But belief in the virtue of work precedes the Puritans. Benedictine monks of the thirteenth century believed that "work is prayer," and that wasting time offended God. They set up elaborate and rigorous schedules designed to thwart the lazy and spur the productive to greater effort. The first mechanical clocks, in fact, were simply automatic bell ringers designed to rouse pious monks from bed and keep them on schedule. Why, in what most people would probably like to imagine as a simpler, more leisurely era, did they have to work so hard?[2]

The answer lies in what they believed about time. In the Judeo-Christian heritage, time has always belonged to God. According to Genesis, God began time by dividing light from dark and setting the heavens moving. And since Adam and Eve's eviction from Eden, God's ownership has demanded hard labor—time on earth must be spent working, to earn our daily bread. If you believe God intended you to work, then it follows that the harder you work, the more you please God. Time in this sense is like a loan from God: men and women have an obligation to use it wisely, to "improve the time," as the Puritans put it. Of course not all Protestants work hard; nor are all hardworking people Protestants.[3] The impulse to work hard is related to the understanding of time, but it's not the same thing.

This is not a study of the pressure to work harder, or a study of how hard people actually worked, but a study of how American ideas about time and its authority changed. It focuses on how we built the web of interconnected, standardized clock time that structures our lives and labor, and how it altered the way we think about ourselves and our society. In the nineteenth century, time changed from a phenomenon rooted in nature and God to an arbitrary, abstract quantity based in machines, in clocks. The development of standard time zones, the mass production of watches and clocks, the invention of instantaneous, synchronized time transmissions and factory punch clocks; all these established new patterns for self-discipline, social order, and the organization of knowledge.[4]

Standard time amounted to a reconstruction of authority—the authority Americans used to govern themselves both in private and public life, at work and in play. By supplanting nature and God with clocks and watches, standard time replaced religious guidelines for dividing up the day, or "natural law," with secular authorities based on efficiency and convenience. Standard clock time, and the mass production of mechanical timekeepers, forced people to reconsider the meaning of machines and mechanical invention—what happened to old notions of individuality, of the individual's place in the scheme of things, when machines became the pattern for social organization instead of nature? Standardized time changed the way Americans thought about themselves and their world.

Not all people experienced these changes in the same way,

because different ethnic and cultural groups approached them with different ideas about time and work. But everyone had to confront them at some point, just as no one living in the United States today, rich or poor, rural or urban, can avoid running up against this framework of standardized clock time. Working, waiting in line, going to a movie, watching TV: all of these demand that we pay attention to the clock, and to an international network built on uniform timekeeping. The history of American time shows how we came to grant the clock such an enormous influence on our lives.[5]

Chapter One describes a small crisis in antebellum American culture, as a sense of time rooted in nature and religion confronted mechanically based sources of time that offered new models for social organization. Chapter Two follows the evolution of a partial answer, regional standard times based on astronomy. These regional standard times helped rationalize interstate commerce and trade. The introduction of national standard time zones by the railroads is the subject of Chapter Three, along with the lasting and virulent hostility they provoked. Chapter Four discusses the mass production and marketing of clocks, watches, and factory punchclocks, exploring the strange interdependence that developed between timekeeping machines and their owners. Standard time established new ways of ordering knowledge, new models for self-discipline. Taylorism, discussed in Chapter Four, provides the transition to the next chapter, which speculates about the connections between time and progressive politics by examining motion picture narrative. The final chapter relates the introduction and repeal of daylight saving during World War I, using the debate to guess at Americans' attitudes toward the framework of standard time created over the preceding fifty years. It ends with the struggle over time, nature, and evolution at the Scopes Trial.

Finishing this book demanded a great deal of ruthlessness,

because a subject as vast as "time" constantly threatens to sprawl out of control. Along the way, friends and fellow scholars suggested many fascinating and potentially rewarding areas of research which I ultimately had to ignore. The book is poorer for it, and the fault is mine. My thanks to the many people, too many to mention all by name, who made working on this project so intriguing. Lawrence W. Levine, Mary Ryan, and Michael P. Rogin, members of my dissertation committee at the University of California, Berkeley, offered a model of generosity and provocative, challenging scholarship. I hope that Professor Levine's commitment to fairness, sympathy, and freedom from clichés echoes in these pages. The graduate students at Berkeley, especially Mark Meigs and Jeff Lena, made it an exciting, friendly intellectual environment, and gave me a chance to play the low post. My thanks also to Anne Hyde, Mary Odem, Lucy Salyer, Nina Silber, and Brian Wiener.

The dissertation that preceded this book was written with the benefit of a Smithsonian Institution Predoctoral Fellowship, 1987–88. The staff, and fellows, of the National Museum of American History made things easier than I had any right to expect, especially Peter Liebhold, Bonnie Lilienfeld, Steven Lubar, Charles McGovern, Jim Roan, and Kay Youngflesh. I owe a very great deal to Carlene Stephens, curator of the Section of Mechanisms at the museum, for her extraordinary generosity in sharing her work, her thoughts, and her time. She exceeds that remarkable Institution's commitment to the diffusion of knowledge, and this project could not have been accomplished as quickly or as well without her help.

A Mellon Postdoctoral Fellowship from New York University, 1988–89, provided time for revisions, while Thomas Bender helped make my stay at NYU both pleasant and productive. Scott Grimaldi let me watch him edit a movie, and Alice Fahs offered useful comments on Chapter Five. Special thanks to Walter Benn Michaels for his perceptive criticisms of a late draft, and to Dan Frank for his skillful editing.

My greatest thanks go to Kimberly Latta, for her critical insights, patience and love.

Contents

Contents

Keeping Watch

I

Time, Nature,

and the Good Citizen

In 1826, New Haven's town fathers paid Eli Terry, Connecticut's most celebrated clockmaker, two hundred dollars to install a clock in the town hall. They wanted a proud symbol of their commitment to order, regular habits, and the virtue of conserving time, and at first the clock served them well. But soon they noticed a growing disagreement with their other source of public time, the Yale College clock. Gradually Terry's clock fell farther and farther behind its rival—five, ten, then fifteen minutes. Perhaps it only needed adjusting. Then it slowly began catching up, raising hopes that it might settle down into steady work as it matured. But instead with each passing day it moved ahead of Yale's timepiece. Finally, almost fifteen minutes faster, Terry's perverse clock began sinking back into its old slothful habits, only to once again start gaining on the faithful college clock, week by week. Had Terry saddled the city fathers with an incompetent timekeeper, or some reckless whimsy?

It seemed unlikely—Terry's reputation for Yankee cleverness

and resolute sobriety had made him famous well beyond Connecticut. Yet there stood the finished clock, proud in its lofty place on the town hall and just as proud in its seemingly bald-faced error. But was Terry's clock wrong at all? Perhaps Yale's clock, faithful for many years, had begun showing its age and running erratically. What if Terry's clock revealed the truth, and Yale's timekeeper only now confessed old habits of duplicity? Then who could say what time it really was?

A letter in the New Haven *Columbian Weekly Register* admitted that the two clocks had become "a subject of much observation and enquiry." But the writer insisted that their disagreement, which "might seem to proceed from negligence," in fact came from two completely different ways of arriving at a definition of time. Yale's clock, designed by the late and well-respected Simeon Jocelyn, followed the sun. Terry's clock offered *mean time*, an average of the sun's daily variation.[1] The letter set off a flurry of heated correspondence, and indeed the difference seems as confusing now as it did then, because Terry's clock posed a difficult but particularly timely question—not what time is it, but what is time?

A bustling market town like New Haven demanded solid answers to both questions. Yale's President Timothy Dwight boasted proudly of New Haven's particular "industry and economy." Busy citizens leavened their business acumen with the college's theological tradition and growing scientific prestige—in 1826, Yale planned on soon establishing one of the first permanent astronomical observatories in America. Native industry, fed by the freshening stream of learning and banked by Protestant religious piety, made his fellows the equal of any Americans in "intelligence, refinement, morals and religion," Dwight happily concluded. He reflected with pleasure on the town's "peace and good order," its freedom from crime and strife, and anticipated a prosperous, peaceful, and well-governed future for his city.

But he also lamented the rapid growth of a class of "*laborers,*" generally "either shiftless, diseased, or vicious," who lacked the sense of order and time-discipline Dwight found so laudable. Their presence cast a darkening cloud over New Haven's industrial future. Dwight's praise of his fellow citizens linked the wise use of time to industrious self-discipline and orderly progress. Yet his sharp criticism of the laboring class, with its apparent lack of industry and indifference to the future, hinted at alternative visions of time and its use, of conflicts to come over both the nature of time and its role in ordering civic affairs.[2] A few years after Dwight's remarks, Eli Terry's clock gave those conflicts substance.

Understanding New Haven's problem requires a brief explanation of the principles of timekeeping. Without really thinking about it, Terry's customers measured time in two different ways. Their simplest measure depended on the sun. When Sol reached the highest point in his passage across the sky, any sundial or even a simple post stuck in the ground read "noon." Anyone could see that, or read the hours on a sundial. But the earth both circles the sun elliptically *and* tilts on its axis. More observant Yankees noticed that the sun's apparent position in the sky varied during the year — noon passed at different points, sometimes higher in the sky, sometimes lower. Astronomers call this annoying habit the sun's *declination*. The variation made no difference if the sun was the only measure of time. When the sundial showed "noon," it was noon — God created the sun to give light and measure time, after all, at the very Genesis of Time itself, and so noon was noon when the sun said so.[3] For obvious reasons, this was called *apparent time*.

But once clocks entered the scene, things got more complicated. Ordinary clockwork cannot follow the sun's irregular movement. Instead, most clocks tell only *mean time*, a convenient average of the sun's apparent variation. Any ordinary, mean time clock set to solar noon on a given day soon begins to drift away from the sun. At its extremes, an ordinary clock read up to sixteen minutes fast or fifteen minutes slow of the sundial four times a year.° Clockmakers had known about the difference between clock and sun time for centuries. They even had a name for it— "the equation of time"—and had devised a few special mechanisms that reproduced the sun's movement. But a clock that showed this "declination" demanded special skills to make and maintain, and cost a lot of money. The vast majority of ordinary clocks told mean time alone, as they always had and still do. When clock and sun varied, most people assumed the clock was wrong, and simply reset it to agree with time by the sun.

Yet the difference between clock and sun was exactly what New Haven confronted, because unbeknownst to most of the town, Jocelyn's Yale College clock had "an apparatus attached to it, invented by himself, which produces a daily variation from [mean time] exactly equal to the variation of the sun."[4] The late Jocelyn's clock, lo and behold, was an *apparent time* clock. It reinforced the astronomy lessons at the college by demonstrating the solar principles of timekeeping, and kept Yale running with the sun. But now it rested in awkward intimacy with an ordinary, mean time clock, with which it would agree only four times a year.

This two-timing arrangement upset New Haven clock watchers. The two clocks questioned the nature of time itself: was it celestial movements or the mechanical timepiece? Signing himself "True Time," one reader of New Haven's *Connecticut Journal* attacked Terry's mechanism. "To have a clock in a town to tell the public what the time is *not* is certainly a novel scheme," he declared sarcastically. "It is said that the clock gives *mean* time," he continued. "But why mean time? Mean time is not true time, nor is true time mean time. A public clock, which tells the truth four

---

° Today, standard time zones and daylight saving have shifted things around, and computing standard clock time from most sundials now requires a whole set of compensating tables.

times only in a year, is something very much like a public nuisance," he concluded.[5] "True Time" demanded to know why the sun—the source of "true" time—had been abandoned for a capricious approximation. Seemingly unaware that almost all clocks could only show mean time, "True Time" echoed a common ambiguity and confusion about time's relationship to clocks.

Clocks were still relatively rare in 1826. Most people, if they owned a clock at all, kept one of Terry's fifteen-dollar "pillar and scroll" shelf models on their mantle. Terry had begun experimenting with interchangeable parts around 1800. By 1826, he had worked out the problems of mass production and was growing wealthy and famous by making and selling ordinary mean time clocks to farmers for miles around. Attractive and affordable, Terry's wooden clocks were not particularly accurate—two or three minutes' variation in a week was optimistic at best. There were also tower clocks, typically situated, like New Haven's, in a church or town hall. These were more accurate, but not by that much; and not always conveniently located, either. The wealthy owned watches, but these also tended to vary, and different timepieces rarely agreed. "Twenty gentlemen in company," almanac maker Nathaniel Low had claimed in 1786, "will hardly be able, by the help of their thirty-guinea watches, to guess within two hours of the true time of night . . . whilst the poor peasant, who never saw a watch, will tell the time to a fraction by the rising and setting of the moon," Low boasted, "which he learns from his almanack."[6] Forty years later clocks were a little more common, but for rural people especially the only reliable indicator of time remained the heavens—the moon and sun's hours of rising and setting as published in the almanac, or the sun's passage shown on a sundial.

"True Time's" letter makes sense, then, if understood by light of the solar timepiece. But it inspired several replies. One "J" explained the difference between mean and apparent time and pointed out "True Time's" mistaken belief that ordinary clocks told apparent time. He insisted that "the public must decide on the comparative utility of these clocks." While suggesting that Jocelyn's apparent time clock might help the faculty at Yale, who conducted their classes by daylight and were concerned to teach astronomy, "J" conceded that most people would simply check their timepieces against a sundial or noon mark occasionally, as they always had. Rather than posing a challenge to New Haven's good order, he explained, the two clockfaces merely demonstrated the scientific principles of timekeeping.

A second, anonymous reply to "True Time," unconcerned by astronomical niceties, insisted that "all the business of our life— our meals—our labors—our hours of rest—everything requires that the day should be divided into equal and uniform portions." It mattered not what standard prevailed, the writer insisted pragmatically, so long as all agreed to use it. Most ordinary clocks and watches could show only mean time—"surely," he reasoned, "the public at large ought not to have all their operations deranged, or their timepieces injured, by attempts to follow the variations of apparent time."[7] But "True Time," even more incensed, dashed off a second letter for the next week's edition.

"Astronomers," he retorted, "have always calculated almanacs, not for *mean* but for *apparent* [solar] time." A clock like Terry's waged "constant war" with the almanac, in fact rendering the almanac nearly useless. "The rising and setting of the sun, of the moon, of the stars . . . and most other celestial appearances are given in every ephemeris in apparent time," he continued, and "all these times are deranged by a clock regulated like the town clock of New Haven." Terry's clock, rumbled "True Time," constituted "an evil of no small magnitude."[8]

The controversy continued over the following months. One correspondent sarcastically wondered if progress might someday enable men to "regulate the sun by the clock." Until that day, he insisted, "we had better be content to go the old way," and, following the college, "regulate our clocks by the sun." He charged that "the citizens of New Haven have been taxed to pay for a clock which gives us false notions of time. Have they not been imposed upon?" If Terry's clock "cannot be so regulated as to keep true time," then "the sooner it is silenced the better. We believe this to be the opinion of MANY," went the conclusion.[9]

In Terry's defense, a reply lamented that "MANY's" remarks had "given uneasiness" to "that worthy artist." It offered the startling claim that Terry's clock "is so constructed as to keep either *mean* or *apparent* time, and also the difference between them." The citizens of New Haven could choose either way of keeping time, "for the clock will do as bid by them." As a young man, Terry had in fact patented an apparent time clock in 1797—the second patent ever issued by the United States. The clock had special gearing and two minute hands of different colors, one for mean and one for apparent time. It had never made him a dime, and so perhaps an inventor's pride had led him to revive the clock for New Haven. At any rate, Terry's champion concluded that when the facts about the new clock "are well known, I trust they will be satisfactory not only to 'many' but to ALL."[10]

But what time would they choose? A final letter, from "A Citizen of the United States," offered a carefully worded description of the difference between apparent and mean time. Although "Citizen" pointed out that in European cities especially, mean time had proven a great convenience, he admitted that New Haven might choose either system. The town's decision remains a mystery, as the correspondence stops here. If Terry's clock could indeed show apparent time, then most likely New Haven kept its public clocks in line with the sun. The clocks could then be regulated by an almanac, or by astronomical observation.[11]

The whole affair might strike the modern reader as silly, but in 1826 there was more to this little incident than the time of day — "True Time" and his opponents debated the nature and source of time itself. The clocks had juxtaposed several concurrent ways of thinking about time. By pitting clock time, a man-made approximation, against the "true" standard of time as revealed in the heavens, and throwing business affairs into disorder, the two clocks dramatized a profound philosophical difference over what time was, who could control it, and how it might be used.

Conflicts like this undoubtedly occurred all over America in the 1820s, because the antebellum decades witnessed a small crisis in the authority of time. As clocks became more and more common, and more and more popular as tools for social organization, Americans began asking these questions more often. Was time a natural phenomenon, rooted in religion and the agricultural traditions of centuries? Or was time an object of rational inquiry, subject to mastery by inquisitive scientists? Or was time simply a convenience, an arbitrary number useful for doing business?

Historians have documented a transition in antebellum America, as an older system of public and private relations, characterized variously as "paternalism" or "republicanism," gave way to new organizations of authority and power. Industrialization and immigration shifted accustomed class relations; new technologies of communication and transportation disrupted market patterns and local economies. The same social and economic pressures that led to religious revivals, to utopian communities, to temperance movements, to reformations in the system of prisons, asylums, and schools, brought with them reorganizations in the systems of time governing labor and public life.[12]

In America at the dawn of the industrial revolution, nature gave evidence of time's passage. The sun's transit, the moon's phases, the passing seasons — all these indicated time and offered a model for using it wisely. In a very real sense, "nature" and "time" meant the same thing. Even the earliest settlers knew about clocks and watches. But they understood mechanical timepieces as mere representations or symbols of time, not as the embodiment of time itself. Where individuals and communities had clocks or watches, they set them to the local sun—each city, town, village, and farm kept its own local time. As the natural world revealed it, time offered Americans lessons in hard work, diligence, thrift, cooperation, rationality, and moral virtue. In almanacs, school-books, and in the patterns of farm work itself, natural phenomena reinforced God's authority and men and women's duty to labor hard and use time, following nature's model, in God's service. But the industrial revolution challenged these assumptions. By the 1830s, a sense of time rooted in nature confronted a seemingly arbitrary time based in commerce, revealing itself in machine movements and the linear progress of invention.

In an agrarian society making the transition to industry, time-keepers posed a difficult and even threatening dilemma. Clocks were first invented to *tell* time, to give a more reliable indication, on cloudy days or at night, of the passage of a quantity belonging to God. The very first mechanical clocks, not surprisingly, simply rang bells to signal the pious monk's hours for prayer. But in the act of telling time the clock tended to become the thing it repre-sented—clocks became not imitations or transcripts of time, but time itself. There was nothing particularly new about clocks gov-erning human movements. In Europe clock bells had been organ-izing town and village life for generations, and the constant tension between "natural" and "mechanical" time they fostered mirrored the conflict of secular and religious authority.[13] But the mass production of clocks—that was new. So too was the factory sys-tem, and the rapid communication and travel it both demanded and encouraged. Nature offered one model for social organization, but as the industrial revolution progressed—bringing with it thousands of Terry's Yankee clocks—the time-telling machine began imposing its own imperatives on life and work.

═══

How did you tell time before the clock? The first and most obvious answer was the sun. Even on cloudy days the sun's passage marked the lapse of time, and on clear days the shadows it cast translated into hours on a sundial or a noon mark inscribed on a wall. In a world less concerned with precise timekeeping, the shadows of trees, rocks, or buildings served the same purpose. Hourglasses measured periods of duration and often timed the minister's sermon, the parishioners perhaps groaning inwardly as the more zealous preachers announced that they would "take another glass." A few tower clocks stood in eighteenth-century America, and the wealthier colonists owned clocks of varying quality, but for people away from major towns the sun, moon, and stars provided the only consistent standard.[14]

On the farm, labor provided its own measures of time. Cows demanded milking twice a day, other animals had to be fed, watered, or sheared; planting, pruning, harvesting—each month, each day brought a round of specific tasks conducted "in their season." Anthropologists, who call this system "task orientation," often view the "natural" time of preindustrial societies as more relaxed, more humane, more respectful of individual autonomy. But for American farm families at least, it is simply wrong to automatically view "task orientation," both in the home and out, as uniformly somehow freer or more leisurely. Farm and house work, spurred by Protestant theology and biblical warnings against idle hands, demanded constant effort. Simply because no mechanical clocks oversaw their activities does not mean that Americans ignored time's passage; in fact, time often presented itself in terms of a sacred duty. Esther Burr, daughter of Puritan divine Jonathan Edwards, frequently complained about the daily grind of her domestic labors. "But I must submit," she wrote in her journal, "my time is not my own but God's."[15]

Two beliefs inform Burr's lament: one about the origin of time, meaning its belonging to God, and the other about the ways time should be used, in her case meaning pious submission to a wearying routine. That she believed time could be possessed is significant in itself; that she believed God's possession of time required her to work hard in obedience to His paternal authority is even more so. We usually understand time in terms of how we use it—the organization of life, the distribution of work and leisure. But this approach obscures the abstract idea of time we use to make our decisions. "Time" must have an origin; it must be considered to have come from somewhere. These origins give time its authority, determining how it should be used. Max Weber's classic analysis suggested that time's origination in God led Calvinists especially to reorganize work and life, as Esther Burr did, around the new imperatives of the Protestant Ethic. Burr needed to "redeem the time," to give evidence of her respect for God's gift—time—by scrupulously saving and using it with piety and care.

What is there of freedom in her obligation to work ceaselessly at turning God's time to profit? What is relaxed about farm and house work's constant round of labor and repair? Where is leisure in a religious creed demanding scrupulous self-improvement even in "leisure" time? Obviously Burr represents an extreme. Lazy people co-exist with busy people in any society, and for every example of industry an example of sloth sits idly by in the shade, laughing. The point here is not to document the amount of work done by any given set of people in a given hour, but rather to explore time's authority—where it comes from, how it is understood, and from this how it can be used to order life. There is a difference between preindustrial time and the understanding of time that replaced it. But the sheer amount of work accomplished measures the difference poorly at best. Often what may look like fundamental differences in attitudes about time in fact have more to do with the kind of work undertaken or the industriousness of the individual.[16]

The difference between Esther Burr's sense of time and our own lies in the symbols and abstractions we use to represent time in public and private life. She found the authority for using her time in God, and saw time revealed in nature; we might find the authority for our decisions about using time in a notion of "efficiency," under the spur of the clock. No precise point of transition from "traditional" to "modern" understandings of time can be fixed; different conceptions of time often co-exist simultaneously in the same society. But understanding the transition between the two requires exploring the idea of time—and the role of the clock—in daily life.

Western European cultures have traditionally understood time both as a circle—a series of repeated natural events like sunrise and sunset or the passing seasons—and as an arrow, shot by God at the dawn of time and stopping only God knows where. Linear and cyclical time have always co-existed, and still co-exist, in our thinking, just as any given society may contain different and even contradictory ideas about time. We depict time, for example, as both a dial, with its endlessly repeated sequence of twelve hours, and as years, totaling an ever higher number toward an uncertain future. Linear time may mean technological and industrial "progress" into that future; it may also recall the inevitability of death and decay. Cyclical time often accompanies the idea of balance in nature, but it just as often recalls the repetitive, clocklike movements of machinery. Our time, in a sense, is like a hoop rolling down a road. In different phases of history and culture we emphasize different aspects of time's duality.[17]

Colonial and antebellum Americans paid remarkably strict and close attention to time's passage and the obligations it imposed. Almanacs were by far the most popular books in the colonies and remained so well after the revolution. American almanacs emphasized the cyclical quality of time, by offering a yearly review of seasonal tasks. But they also served as guides for understanding, interpreting, and managing time correctly as it rolled along toward God's eternity. The second book published in America was an almanac, at Cambridge in 1638 or 1639, and the Cambridge press dominated the field for its first thirty-five years. Products of Puritan Harvard, these early "Philomath" or astronomical almanacs concentrated on explaining the phenomena of the heavens in terms of God's providence, willfully overlooking any contradiction between science and religion that might suggest itself. Typically they combined the two, supplying the times of the holy Sabbath and helping clarify the mysteries of God as revealed in the heavens and the natural world.[18]

Almanac maker Samuel Atkins, traveling through Maryland in 1686, declared that the people "scarcely knew how the time passed, nor that they hardly knew the day of rest, or the Lord's Day, when it was, for want of . . . an Almanack." Without a guide to time reckoning Marylanders simply could not organize their days properly, he claimed. "On the other side," Atkins continued, "having met with Ingenious persons, that have been lovers of the Mathematical Arts, some of which have wanted an Ephemeris to make some Practice thereon," he decided to offer his own almanac "to these my Country men."[19] Atkins's solar and lunar timetables merged Christian piety, science, and the practice of farming. As a guide to holy days his almanacs enabled people to orient themselves within God's time and respect sacred law—to observe the "correct" times for work and worship—while his precise measurement of the heavens' movements satisfied and encouraged scientific fascination with God's creation.

After about 1680, almanacs began to reflect less concern with theological orthodoxy and more with worldly affairs. These newer almanacs increasingly blended Copernican science with the flavors of deism, as the following poem from "Abraham Weatherwise's" 1759 Almanac demonstrates:

On A Watch

Could but our Tempers move like this Machine,
Not urged by Passion, nor delay'd by Spleen;
But true to Nature's regulating pow'r,
By virtuous Acts distinguish'd every Hour;
Then Health and Joy would follow as they ought,
The Laws of Motion, and the Laws of Thought:
Sweet Health, to pass the Laws of Moments o'er,
And everlasting Joy, when Time shall be no more.

"Weatherwise" borrowed from an Enlightenment tradition which used the watch as analogue to God's creation. The movements of the heavens indicated time, as did the watch; they operated according to fixed and discernible principles, like the watch, and the poem suggests that we model our behavior after these movements. The regularity of the physical world as it marks time should be our pattern, and since we can discover laws within its motions, it follows that there must be laws governing our "Tempers"— "Laws of Motion" as well as "Laws of Thought." The well-tempered farm, like the well-tempered society, imitated the regular and clocklike movements of nature, the source and measure of time.[20]

But the clock, in this passage, never represents the *source* of time. Rather, it serves only as an embodiment of human under-standing—of our ability to comprehend nature and natural laws, and through them the passage of time. The watch is "true to nature's regulating power": nature, not the mechanical device, is the model for the organization of society, and time, in this typical instance, is a by-product of nature. Nature's regular and predict-able movements provide clues to the mind of God and thus to the proper management of time as God intended it. Almanacs charted these natural movements—which were understood by readers as Time itself—and helped interpret them.

Although colonial almanacs included much more, solar and lunar timetables, together called the *ephemeris*, were their main attraction. The ephemeris usually offered the precise hours and minutes of the sun's rising, the moon's southing, high and low tides, eclipses, and the movements of the major stars. Typically they devoted one page to each month, and combined the times and dates of celestial appearances with astrological symbols for interpreting them, weather forecasts, historical information, and miscellaneous advice to farmers.

The celestial rationalism evident in the poem quoted above blended easily with what we would call superstition. Eighteenth-century almanacs usually included a drawing, called the "anatomy" or "the man of signs," of the human body marked by the signs of the Zodiac. Each zodiacal symbol controlled or affected a specific part of the body. By consulting the anatomy, one could learn the best time for bleeding a diseased arm or leg, or conceiving a child. Even more commonly, the stars offered advice on farming—when to plant, wean animals, mend fences, and so on. Cotton Mather, for example, believed that timber cut in the waning moon of Au-gust would never suffer from worms. Like occultism and folk astrology, the practice of reading the stars for advice on farming continued well into the twentieth century.[21]

The almanac offered a guide to the stars, helping users understand and manage their lives by situating them in time. Within its pages astrology and astronomy, theology and husbandry, medicine and science combined to schedule work. Readers went to the almanac's timetables for help in managing these different categories of knowledge—on the one hand to discover what time it was at that moment, and on the other to discover the most appropriate time for doing some task. The connection of time to nature, and thus to God, that almanacs reinforced points to an American obsession with time, its measurement, and its proper use.

The most successful almanacs made time management their explicit focus, and of these Robert Bailey Thomas's *Old Farmer's Almanac*, by the early nineteenth century, had become the model for imitators all over the country. Thomas added practical and homely advice to the ephemeris while downplaying astrology and homeopathy somewhat. His "Farmer's Calender" specialized in cautionary messages—for the first week of October 1800 the calendar urged: "Winter apples should now be gathered up, as the frost hurts them much . . . harvest your Indian corn without delay—the birds and squirrels I am confident will."[22] Thomas reminded dilatory readers that each task had its season, and its season was short. But much of the material Thomas printed, like the pages reproduced in part below, now seems nearly indecipherable—such common familiarity with the sun and sky has largely passed away.

Tables like these suggested that a prosperous, orderly, and well-tempered life could be gained by following nature, by running life according to the clock in the sky. They make little distinction between different kinds of events—Bonaparte's death in 1821 (bottom panel, line six) has the same significance as the quadrature of the Sun and the planet Uranus. Though dropping the "man of signs," this most successful of almanacs continued charting the heavens' effects on the human body—column nine of the

Blossoms scatt'red o'er the trees—
Odours wafted in the breeze
Is the bliss of every day,
Towards the last of blooming May.

S.M.	S.W.	Courts, Aspects, Holydays, Weather, &c. &c	FARMER'S CALENDAR.
1	7	St. Phil. & St. Ja. ☐ ☾ ☿	Farmers should be as busy as Bees.
2	C	3d Sun. past Ea. *Fine*	See now, while the flowers are in
3	2	Mid *weather.*	blooming, the inhabitants of the hive
4	3	[Gen. El. R. I. & Con.	are all in motion! The laborers take
5	4	St. Jn. Ev. ♃ Sta. ☽ Apo.	wing and go abroad in search of hon-
6	5	tide. [Bonap. d. '21. ag. 52.	ey and wax. Honey is a limpid juice
7	6	*The weather*	found in the blossoms. Wax is made
8	7	*continues* ☽ ☉ ♂	by the bees from the dust contained
9	C	4th Sun. past Ea. *fine.*	within the blossoms. These different
10	2	*More* [Con.	materials being brought to the hive,
11	3	S. J. C. Len. & Ply. C.C.	the laborers in waiting take the wax,
12	4	*unsettled for some*	and form of it those little six-sided
13	5	Jup. sou. 4h. mor. *days.*	cells, which serve as storehouses for
14	6	*High winds.*	the honey, or nests for the young. The
15	7	Spica so. 9h. 45m.	honey is partly distributed for present
16	C	5th Sun. past E. ♀ G. El.	food to the inhabitants, and the re-
17	2	C. P. Edg. Quite *Rain.*	mainder laid up for winter. The
18	3	[D. Dav '80 ☽ ☽ ♀	queen bee now begins to deposit her
19	4	St. Duns. C. C. Edg. ☽ per.	eggs, about 50 each day, in the empty
20	5	Ascen. H. Thur. △ ♀ ♄	cells. The egg, being soon hatched
21	6	high *More*	into a little grub, increases the em-
22	7	tides. *showers.*	ployment of the laborers, whose care
23	C	Sun. past Ascen.	it is to feed it with purest honey.
24	2	C. P. Nant. *Cooler.*	When it is full grown, its cell is seal-
25	3	*Fine settled*	ed up with wax. Its form then alters,
26	4	Gen. El. Boston. *weather.*	and in a few days it breaks its prison,
27	5	Jup. sou. 3h. mor.	being changed into a perfect bee, quits
28	6	Low *Signs of*	the hive, and, joining in the general
29	7	tides. *rain*	employment, goes in search of honey
30	C	Whit-Sun. *Very warm.*	for the public store. Let man take
31	2		a lesson from this little economizer.

"Be cheerful, social, friendly, kind,
Such as may improve the mind;
Where health and temperance may join,
And Virtue own the soft design."

Pages from Robert Bailey Thomas, The Farmer's Almanac, *May 1830 (Boston, 1829).* Thomas began calling it The Old Farmer's Almanac *in 1832, to distinguish himself from imitators.*

1830. MAY hath 31 Days.

ASTRONOMICAL CALCULATIONS.

☉ Declination.	Dys.	d.	m.	Dys.	d.	m.	Dvs.	d.	m.	Dys.	d.	m.	Dys.	d.	m.
	1	15 N	0	7	16	45	13	18	20	19	19	48	25	20	55
	2	15	18	8	17	1	14	18	34	20	19	56	26	21	5
	3	15	36	9	17	18	15	18	49	21	20	8	27	21	16
	4	15	54	10	17	34	16	19	3	22	20	20	28	21	26
	5	16	11	11	17	49	17	19	17	23	20	32	29	21	35
	6	16	18	12	18	4	18	19	30	24	20	44	30	21	44

● Full Moon, 7th day, 7h. 18m. evening.
☾ Last Quarter, 15th day, 11h. 34m. morning.
○ New Moon, 22d day, 2h. 29m. morning.
☽ First Quarter, 29th day, 6h. 4m. morning.

D.M.	W. D.	☉ r.	☉ s.	l. r.	D. rise	D. mo.	☉ F. sea. A.H. m.	☽'s place	r. & s.	☽ sou.
1	Satur.	5 3	7 13	51	5 0	3 9	7 6	heart	2 7	7 56
2	SUN.	5 2	7 13	56	5 2	3 10	7 52	belly	2 49	8 40
3	Mond.	5 1	7 13	58	5 4	3 11	8 37	belly	3 10	9 23
4	Tuesd.	4 59	8 14	2	5 8	3 12	9 20	reins	3 38	10 5
5	Wedn.	4 58	8 14	4	5 10	3 13	10 3	reins	4 6	10 47
6	Thurs.	4 57	8 14	6	5 12	4 14	10 49	secrets	4 56	11 31
7	Friday	4 56	8 14	8	5 14	4 ● 11	34	secrets	● rises	morn.
8	Satur.	4 55	8 14	10	5 16	4 16	morn.	secrets	7 52	0 15
9	SUN.	4 54	8 14	12	5 13	4 17	0 20	thighs	8 46	1 2
10	Mond.	4 53	8 14	14	5 20	4 18	1 5	thighs	9 40	1 49
11	Tuesd.	4 51	8 14	18	5 24	4 19	1 53	knees	10 30	2 39
12	Wedn.	4 50	8 14	20	5 26	4 20	2 41	knees	11 17	3 29
13	Thurs.	4 49	8 14	22	5 28	4 21	3 31	knees	morn.	4 21
14	Friday	4 48	8 14	24	5 30	4 22	4 21	legs	0 1	5 13
15	Satur.	4 47	8 14	26	5 32	4 23	5 12	legs	0 40	6 4
16	SUN.	4 46	8 14	28	5 34	4 24	6 6	feet	1 17	6 56
17	Mond.	4 45	8 14	30	5 36	4 25	7 0	feet	1 53	7 48
18	Tuesd.	4 44	8 14	32	5 38	4 26	7 55	head	2 28	8 41
19	Wedn.	4 43	8 14	34	5 40	4 27	8 50	head	3 3	9 35
20	Thurs.	4 42	8 14	36	5 42	4 28	9 46	neck	3 41	10 30
21	Friday	4 42	8 14	36	5 42	4 29	10 45	neck	4 22	11 27
22	Satur.	4 41	8 14	38	5 44	4 ○ 11	42	arms	☽ sets.	eve. 26
23	SUN.	4 40	8 14	40	5 46	4 eve.	39	arms	8 49	1 25
24	Mond.	4 39	8 14	42	5 48	4 2	1 36	breast	9 49	2 24
25	Tuesd.	4 38	8 14	44	5 50	3 3	2 30	breast	10 42	3 20
26	Wedn.	4 37	8 14	46	5 52	3 4	3 22	heart	11 28	4 14
27	Thurs.	4 37	8 14	46	5 52	3 5	4 10	heart	morn.	5 4
28	Friday	4 36	8 14	48	5 54	3 6	4 56	heart	0 8	5 51
29	Satur.	4 35	8 14	50	5 56	3 7	5 41	belly	0 40	6 36
30	SUN.	4 34	8 14	52	5 58	3 8	6 26	belly	1 12	7 20
31	Mond.	4 34	8 14	52	5 58	3 9	7 10	reins	1 41	8 2

"Astronomical Calculations," for example, indicates the moon's effect on heart, belly, reins (ribs), and "secrets," respectively. These astrological remnants reinforced the idea of men's and women's connection to cosmological mechanism, suggesting that just as in a watch each part affects the other, so we too are affected equally by the parts of the physical world.

And by connecting seasonal labor to celestial motion, the almanacs made it possible to regulate work and find the best way to do it—if the signs were read correctly, or nature's example was understood properly. But again, this "natural" time should not be mistaken for a more leisurely approach to work. "System, Mr. Hasty, system," the almanac warned in 1832. "Your plans for the whole season should be well laid out and adjusted, in doing which, you should take some account of foul weather as well as fair."[23] *The Old Farmer's Almanac* supplied the system, using nature as its regulating authority. The quotation above ended by reminding the farmer that time is money, and that "system"—organizing time by natural signs—helped enormously in straining the most value from hired help.

American almanacs, by the 1830s, reflect a transition in thinking about the ends of work; natural examples begin emphasizing work for money rather than work for work's own sake. Samuel Atkins's pious and scientific fascination with the stars merged with the profit motive in the almanacs of these years. But though *The Old Farmer's Almanac* of the 1830s reflected a close connection to market economies—it included tables of interest, mileage between principal cities, and the hours of county, state, and federal courts— it insisted on nature as the pattern for success in life and the guide to time usage. The "Farmer's Calender" reinforced the point by comparing the farm family to a hive of busy bees—while the flowers bloom, the inhabitants of the farm-hive, "the laborers," are up in motion. While the sun, moon, and stars marked time, the prosperous farm family looked to nature for living models of diligence, thrift, and hard work.

"Pretty bee, will you come and play with me?" asked "the Idle Boy" in an early *McGuffey's Reader.* But the serious-minded bee replied, "No, I must not be idle, I must go and gather honey." Rebuffed, the idle boy asked a dog, a bird, and a horse in turn, but each was righteously busy. "What, is nobody idle?" the boy asked in a moment of revelation. "Then little boys must not be idle." So he "made haste, and went to school, and learned his lesson very well."[24]

Like almanacs, nineteenth-century schoolbooks linked time to nature and farm work. But they nagged even more about the morality of working hard—"the industry of the ant and the bee and the fecklessness of the grasshopper are ever present."[25] Primers and readers emphasized the sin of wasting time, hinting at disasters waiting in ambush for all those who stopped to smell the roses. "Little girl," one *McGuffey's Reader* warned, "never be a moment too late. It will soon end in trouble or crime."[26]

What sort of crime was idleness? "Time," wrote educator Lyman Cobb in 1843, "we ought to consider as a sort of sacred trust committed to us by God; of which we are now the depositories, and are to render an account at the last." Cobb offered a variation on the old Puritan phrase "improve the time." Time was a loan from God, an investment made in men and women, and God expected some return, some increase in spiritual (or worldly) profit. Such a gift could neither be wasted nor taken lightly; it required the same regular ordering seen in God's example, nature. "Where no plan is laid," Cobb continued, "where the disposal of time is surrendered merely to the chance of incidents, all things lie huddled together in one chaos, which admits neither distribution nor review." Idleness, and failure to adopt nature's model, profaned God's intentions and opened the door to chaos.

Cobb's little homily suggested that "system," or the organization of time, on the one hand ensured God's favor and, on the other, made the individual's time both usable and useful to society—subject to "distribution and review." "The orderly arrangement of time," Cobb continued, "is like a ray of light, which darts itself through all affairs."[27] Temporal discipline, Cobb implied, is the road along which knowledge travels, and his essay suggested that organized time served as the die that stamped good citizens out of formless material. The ubiquitous textbook example of the bee and the ant, both social insects, reinforced this message of organization and control. Rigorous organization of individual time made for an organized and rational society. And as an organizing principle, time rooted in orderly nature illuminated the dark corners of society and put the loafing minions of "chaos" to useful work.

The idea of organized time as an illuminating principle recurred frequently when time was related to nature, and especially the sun. "Behold the sun in his splendor moving over you by day; and the moon at night," suggested one 1810 reader. "Contemplate a steady and powerful Hand, bringing round spring and summer, autumn and winter, in regular course." The sun and moon gave evidence of God in action, and illuminated God's plan for men and women. "Can you believe," the lesson asked, "that you were brought hither for no purpose, but to immerse yourselves in gross and brutal, or at best, in trifling pleasures?"[28] The march of time, revealed by blazing sunlight in the glory of nature, encouraged imitation by God's foot soldiers on earth.

In its emphasis on system and relentless hard work this rhetoric may now remind us less of a grand march than a lockstep— it sounds a bit like "scientific management," the late nineteenth-century credo that treated workers like parts in a clock/machine. But in these lessons, time came from nature, not clocks. Time's authority demanded systematic work, but the authority submitted to, the steady and powerful hand, belonged to God. The use of time remained tied to local, natural imperatives. Cyclical nature embodied time, not clocks, and nature recalled the power of tradition and the moral virtue of a productive, well-ordered society of farmers. One reader, as late as 1857, still explained the "division of time" in terms of the seasons and their labor rather than as hours, minutes, or seconds. "In September, October, and November the produce of the fields is gathered. The farmer cuts his grain. . . . It is put in the barn and threshed." The lesson ended by pointing out that it is now more than 1850 years since He [Christ] came on earth."[29] One thousand eight hundred and fifty years in the unfolding of God's design, ticked off in the circle of passing seasons—time, God, nature, moral virtue, and economic prosperity were inextricably linked.

Schoolbooks ignored the lilies of the field, never known for their industry, in favor of the tireless bee or ant's example of diligence and frugal, orderly social organization. "Nature [was] never neutral" in children's texts.[30] In its relentless glorification of hard work, the "natural time" these readers and almanacs set forth hardly seems leisurely. The lessons alone, of course, fail to tell us how hard antebellum Americans actually worked or would work. Most reflect the religious and class biases of the New Englanders who wrote them. Schoolbook homilies on the busy bee may have amounted to little more than a nagging buzz in the ears on "Saint Monday," the artisan's traditional leisurely day of recovery from weekend revels. But schoolbooks and almanacs do tell us about how differently time was once understood.

The crucial issue, again, is time's perceived source. Time belonged to God—and God demanded hard work—but time revealed itself in nature, which showed in turn how it should be used. "Time is measured by clocks and watches, dials and glasses," Noah Webster admitted. But "the light of the sun makes the day, and the shade of the earth makes the night . . . the day time is for labor, the night for sleep and repose." "Children," he added for good measure, "should go to bed early."[31] Natural cycles still modeled the child's habits in this passage. But Webster's description hints at a crucial transition to come—a transition from time rooted in nature to time originating in the clock itself. To explore this transition, we need first to examine the ambiguous place of clocks in antebellum American culture.

Settlers brought clocks with them to the colonies as far back as 1638. Boston had a town clock, and a man assigned to tend it, in 1668, and clockmakers plied their trade in Philadelphia in the 1680s. There were probably clockmakers in every colony by 1700. Advertisements for tower clocks appeared in Boston's first newspaper, and tower clocks themselves in Connecticut in 1727.[32] These earliest tower and house clocks were relatively crude affairs, technologically primitive by the standards of the Old World. But though no substantial domestic watch industry appeared before the Civil War, American clock production soon rivaled Europe and England in quantity at least. It would be interesting to know who wanted these early clocks, how they were paid for, and who owned them, but very little research on the subject exists. The best estimates to date suggest that by 1700, roughly one white adult in fifty kept a clock, and one in thirty-two a watch, and that among this minority merchants, professionals, innkeepers, and shopkeepers predominated. Artisans were twice as likely as farmers to have the means of telling time, and on an average "almost half of the men who lived in major cities had a timepiece of some kind."[33]

In the city a different kind of time prevailed. Since the Middle Ages and probably earlier, city life had made special demands on people's time.[34] The hours of markets needed attending to, artisans and merchants needed to assemble at a given time, courts and government offices had their schedules of operation; plays and festivals had to begin at a certain time. On the surface, this busy round looks like the city life of today. But the eighteenth and early nineteenth centuries paid considerably less attention to punctuality and saving time than ourselves. Some evidence, for example, suggests that colonial businessmen worked at a far more leisurely tempo than today's. The starting times for plays, public festivals, or political meetings remained loosely defined—and loosely observed, by our standards—until the late nineteenth century. And antebellum workers cheerfully ignored the more structured hours of clock time whenever they could get away with it.[35]

But life in cities and larger towns did make unique demands on people. It involved them in each other's lives and work, requiring agreement about when to meet. And since work and leisure in the city often continued year round, rain or shine, only mechanical timekeepers allowed business as usual on cloudy days, or at night. In most communities a town clock, the sort of prominent tower clock New Haven's city fathers had envisioned for their town, filled the bill. In New England, town clocks usually appeared first on churches, paid for by public subscription. Such clocks, placed on the steeple, connected clock time to God while tying social organization to religious authority—founded solidly on the Bible and the Church, these clocks stood above the dirty business of industry and commerce. Later, as town clocks appeared on public buildings other than churches, they symbolized order and the impartial regulation of daily life.

Despite their use as public timekeepers and their technological sophistication, these sort of town clocks rarely acquired the negative connotations attached to other equally complex machines, like the railroad or cotton loom. They conveyed a time sanctified by religion and tradition and thus "natural." As late as the 1840s the town clock's summons to Sunday service remained a set piece in pastoral writing, recalling the orderly, Godly life of the rural past. A good example appeared in Hawthorne's 1844 *American Notebooks*, when the retiring author described a locomotive's rude desecration of the woods. Before the locomotive arrived to disturb his reverie, Hawthorne listed a variety of the soft and pleasant sounds he heard in the woods, including the striking of the village clock. The ringing village clock, in fact a bravura technological achievement and a symbol of social regulation, is passed over as part of the pastoral scene that precedes the train's arrival—akin to the natural sounds of farm labor and totally unlike the hellish locomotive. More than simple timekeepers, in this pastoral guise mechanical clocks represented the principles of natural law and their applicability to a virtuous, orderly local government.[36]

An age that often compared God to a watchmaker and the universe to a clock, like "Abraham Weatherwise" in his almanac, regarded clockmakers as much more than artisans, and attached far more than local importance to their work. The trade required a great deal of mathematics and astronomy, and the best American clockmakers claimed equality with philosophers and scientists. David Rittenhouse, the foremost example, rivaled Franklin as a leading thinker among his peers. Besides establishing the first astronomical observatory in America, Rittenhouse also produced elaborate, extraordinarily well-crafted clocks depicting the movements of the moon, stars, and planets along with hours, minutes, and seconds. His crowning mechanical achievement was an *orrery*, or clockwork model of the solar system, made for the University of Pennsylvania. The orrery inspired Jefferson to claim that Rittenhouse "has not indeed made a world, but by imitation approached nearer its Maker than any man who has lived from the creation to this day."[37] The orrery emphasized, like almanacs and schoolbooks, time's cyclical aspect, and the orderly beauty of nature's movements.

Rittenhouse created a machine that imitated and displayed natural law; his clock revealed not just the time of day but time's

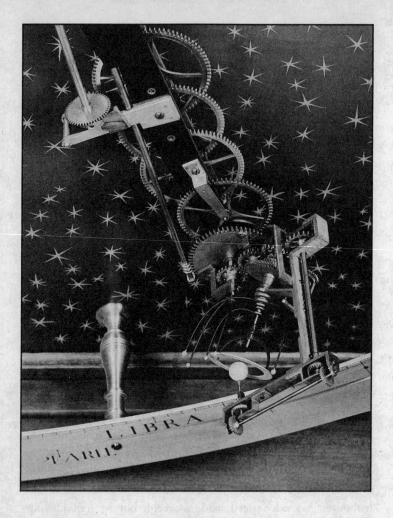

The Saturn arm of David Rittenhouse's orrery, 1767–71.
One of two very similar orreries built by Rittenhouse, now in the
Van Pelt Library of the University of Pennsylvania. The orrery
dramatizes the era's connection of mechanisms to the
"clocklike" workings of natural law. (National Museum of
American History, Smithsonian Institution.)

celestial cycle itself, as well as it could be understood. Inspired in part by the model of a clockwork universe, American political thinkers interpreted these celestial movements as a pattern for government. Federalist political theorists often compared the Constitution to a clock/machine, "a machine that would go of itself," like the deist's perfect watch. Like the seasons, or the mechanical forces that governed planetary movement, the Constitution was supposed by some to be "timeless," a permanent restraint on democracy's inevitable and universally assumed tendency to decline into tyranny over linear time. With its system of "checks and balances" on the mechanical forces of human nature, "the Federal Constitution was . . . a pendulum clock in perfect balance." As a rational machine based on scientific principles, the Constitution "governed" natural political forces like the safety devices on a steam engine.[38] Just as the village clock preserved the orderly virtue of the pastoral village, so the clockwork model promised to order politics and preserve stability.

But two potential interpretations of the clock/machine vie for supremacy in this understanding of the Constitution. One locates political authority in imitation of natural law, while the other implies supplanting natural law with human ingenuity—as Rittenhouse's machine makes him literally godlike, so too the achievement of the convention brings the delegates, as creators themselves, closer to God. Thus the Constitution represents not just stability but *perfectibility*, the virtue of progress and enlightened reform. The two interpretations echo the tension between cyclical and linear time, between stasis and progress. The Constitution could be seen as a regulating machine, maintaining regularity and stability. But it might also be seen as a sort of throttle on the speed of progress. If government, like Rittenhouse's orrery, could be brought ever closer to perfection, then there need be no end to human tinkering with government, society, and technology—no end to "progress."

The tension inherent in these two interpretations, and the idea of time that informed them, reappeared in the conflict between Jefferson and Hamilton and their respective political camps. On the farm, Jefferson passionately maintained, life followed nature and the seasonal cycle. Natural law governed farming, assuring its virtue. A society of Jefferson's "yeoman farmers," almanacs in hand, was a "naturally" virtuous society. Associating industry with corruption and dependence, Jefferson hoped, not without skepticism, that the agrarian republic might preserve itself intact by expanding geographically across western space, rather than through time in the ore-car of industrial progress. It is true that Jefferson, an inventor himself, admitted no objection to home manufactures and local, cottage industry. Such small-scale manifestations of technological cleverness stayed within the framework of cyclical time and natural imperatives. But he feared the centralizing tendencies of large-scale manufacturing and the inevitable formation of a laboring class to support it — England's Manchester slums haunted his dreams. Jefferson's generation interpreted the "fall" of classical Greece and Rome as the collapse of agrarian democratic virtue. For Jefferson, progress through time presaged American democracy's senility and eventual death.[39]

But where Jefferson saw possibilities for decline, Hamilton dreamed of progress. Democracies may have collapsed repeatedly in the past, but the Enlightenment hinted at the eventual perfection of philosophy and technology — in a climate of political stability, America's future held fantastic promises of industrial progress and wealth. Hamilton hoped to follow the path of time's arrow and set the new states on course into an industrial future. Jefferson's system "sustained a vision of social and economic development across space, whereas the other system [Hamilton's] supported an alternative vision of development through time."[40]

Supporters of Hamilton's program of manufactures, as Leo Marx has pointed out, were forced to drape their arguments in pastoral sheep's clothing; Tench Coxe, for example, used Rittenhouse's orrery to promote machine power and industry. Coxe "enlisted the immense prestige of Newtonian mechanics in support of his economic program" while directing society forward into industrial development. He tried to claim that textile machinery, like the orrery, embodied natural laws. In this sense the very perfection of Rittenhouse's orrery, like the natural laws embodied in the Constitution, could lay fears about the corrupting power of machinery to rest. Having discovered the principles of rational government and learned how to reproduce them, society was free to advance into the Hamiltonian future of industrial improvements.[41]

The career of John Fitch, inventor of the steamboat, offers another example of the tension between cyclical and linear time and the ambiguity inherent in the clockwork metaphor. Fitch, a Connecticut Yankee, began his career as a clockmaker's apprentice. The craft gained him familiarity with metals, tools, and the techniques of working them, as well as insight into the harmonious movements and principles of Newtonian mechanics. But his pious, Puritan master's disregard for profit, pleasure, and advancement, his insistence on limiting both mechanical innovation and business opportunity, left Fitch frustrated. Rejecting both theological orthodoxy and the clockmaker's trade, Fitch embraced a flabby deism as he turned his attention toward the steam engine. The steam engine satisfied his mechanical predilections while deist philosophy, assuring the rationality of his actions and his machines, allowed him to vault the restricting theology of his forebears and set his course into the linear time of the industrial future.[42] Fitch's famous invention mirrored changing ideas about time and clocks: it converted the cyclical, clocklike motion of the steam engine's parts into linear progress through the water and through time. Only by rethinking his understandings of cyclical and linear time—and turning his clockmaker's skills to more innovative ends—could Fitch free himself from tradition and liberate his mechanical creativity.

Mass-produced, affordable clocks and watches exacerbated the tension between industrial, linear time, and the cyclical time of nature. When tied to nature, as in Rittenhouse's orrery and the pastoral village clock, clocks represented cyclical time. They recalled the governing mechanisms God had established and God's children discovered in natural law. A clock regulated by the almanac, as "True Time" recommended, humbly imitated God's time. On the other hand, mechanical clocks kept working when the shy or modest sun draped himself in clouds. Their value in organizing city life and labor suggested that mechanical creations had improved on nature's timepiece.

The progress of industry and commerce itself depended on clocks and clock time. If a mechanic, out of earshot of the town clock, needed to leave home for work by noon, owning a clock made his task easier on a cloudy or rainy day. If men worked under him, the good fortune of owning a watch gave him power, since he could now call the hours of work and rest if no other clock stood nearby. The watch itself, in the latter instance, becomes the authority for time, and its owner derives power from ownership. Life in the cities and market towns demanded manmade, mechanical sources of time, sources of time that could be possessed and controlled.

If the almanacs and schoolbooks quoted previously established time's connection to nature, these city clocks threatened to overturn nature and substitute human authority for God's. Recalling the incident in New Haven that began this chapter, we can see why "True Time" was so agitated. As machines, clocks could stand for either cyclical or linear time. On the one hand, clock metaphors helped make the universe both comprehensible and reassuring while offering a model for the justly organized society—in its pastoral guise the clock illustrated cyclical time and the ideals of Jeffersonian republicanism. On the other hand, clocks also represented human cleverness, commercial improvement, and progress—the Hamiltonian vision of an industrial future. "True Time" called Terry's clock "an evil of no small magnitude" because by exploring the nature of time it echoed lasting debates over the foundations of political and moral authority. Was society to be regulated by nature and God, or by the clock, with its associations of industry and the ideal of progress?

Industrial progress demanded clocks, and ironically the first application of interchangeable parts to manufacturing came at Eli Terry's Connecticut clock factory. Terry had served a traditional apprenticeship and then run a small shop making tall case clocks for several years. But in 1806 he began producing clocks on a larger scale, converting an old textile mill to a clock factory and using a small water wheel to turn gear-cutting engines. By 1822, after a series of production and design innovations, Terry had perfected his most lasting and lucrative product, the "pillar and scroll" shelf clock made almost entirely of wood. Its relatively small size made it easy for peddlers to carry while its simple, attractive design made it easy to sell and its familiar materials easy to repair. The clock "completely revolutionized the whole business," and set the standard for an industry.[43]

Most clockmakers had waited for a customer's order, then laboriously crafted a clock by hand. Terry assumed the market already existed—he either guessed that demand for clocks was increasing or assumed aggressive sales would move more clocks, or perhaps both. The "pillar and scroll" clock quickly became associated with the archetypical fast-talking Yankee peddler, the glib salesman of wooden nutmegs.[44] Terry's great success initiated a long period of intense competition, innovation, and "progress" in the design and marketing of clocks. Though he filed a barrage

*Eli Terry "pillar and scroll"
clock, ca. 1820. This particular
clock is somewhat unusual in
having the escapement and pendulum rod in front of the dial
plate rather than hidden behind.*
(National Museum of
American History,
Smithsonian Institution.)

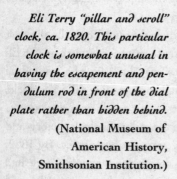

of patents, Terry's clock was widely copied. Clockmaking boomed
in Connecticut, and in 1836, the peak year, more than 80,000
Terry-styled clocks sold for ten dollars or less.[45]

Terry's clocks, and later, the cheaper brass clocks Chauncey
Jérome and others sold for as little as $1.50, brought mechanical
timekeeping within the easy reach of mechanics, farmers, artisans,
and laborers for the first time. They sold extremely well in every
state. One traveler claimed in the 1840s that "in Kentucky, in
Illinois, in Missouri, and here in every dell in Arkansas," and even
"in cabins where there was not a chair to sit on, there was sure
to be a Connecticut clock."[46] By 1835 Yankee clocks were so
popular in the South that a number of states, hoping to promote
manufactures at home, raised the clock peddler's license fees to
discourage their sale. Chauncey Jerome's firm responded by shipping disassembled clocks to Richmond, Virginia, and Hamburg,

South Carolina. Assembled on southern soil, Jerome's popular clocks satisfied the demand for local manufactures.[47]

Associating clocks as we do with punctuality and industrial labor, we tend to assume such popularity reflects the spread of commercial markets and industrial capitalism—the linear time of industry. But a closer look at the clocks suggests that such an interpretation is too simple. Their design, for example, tended toward the austere, mimicking traditional furniture designs and eschewing the whimsical, consumerist frivolity that would characterize clock styles after the Civil War. One of Jerome's most popular designs, the "Gothic" shelf clock, looked remarkably like a New England church. Simple and even grave in appearance, Terry's clocks called attention to the seriousness of time.

Most Terry clocks included a small painting, called a "tablet," beneath the face. The tablets often depicted a mill or factory, and indeed the clocks were used to regulate manufacturing.[48] But they more typically depicted farm life, the rural village or conventional landscape, firmly connecting time to nature and the pastoral. The pendulum, clearly visible through the opening in the tablet, recalled the pulse of time in nature's cycle and the balanced mechanism of the natural world as it operated in daily life. Such a clock advertised its owner's piety and respect for God's time.

While the ticking clock may have been connected to cyclical nature in Terry's designs, a substantial body of American folklore linked clocks with mortality and the linear brevity of life. Their striking, coming unexpectedly, was said to denote death. A mysterious ticking, heard where no clock stood, was commonly referred to as the "wall clock" or "death watch." Recalling "the ticking of an old-fashioned clock, reckoning time towards eternity," this superstition appears as a portent of death in the folklore of at least thirteen states.[49] A suddenly silent clock indicated death or at least very bad luck. "A clock stops when someone dies . . . when there is a death in the family. . . . If a clock suddenly stops, it's a sure sign of death."[50] The clock, in these examples, embodies God's time—a silent clock showed that death had brought someone's time to a halt.

Altering the time, or allowing a clock to run down, was therefore dangerous: "never turn a clock counterclockwise . . . a run-down clock is a sure sign of death." Other proverbs held that broken clocks should be buried or stored well out of sight, and that one should "never keep a clock that does not run . . . it is bad luck to keep a clock that is not running."[51] The uneasiness stemmed from the fact that though they measured God's time, clocks were also easily susceptible to human control. Adjustable at the owner's whim, they threatened to present an alternative source for time, a secular time made and controlled by men and women. Time's measurement, and its authority, were not to be trivialized—keeping a broken clock suggested an indifference to God and God's time, much as gambling flouted God's control over earthly events.

The repeated appearance of a belief that "it's bad luck to have two clocks running in the same room . . . two clocks running in the same house" further attests to this anxiety. "Never leave two clocks ticking in the same room at the same time, very bad luck," one folklorist's informant claimed. "My Grandmother would not do that for anything, for she said it was a sure sign of death."[52] No two clocks can be made to keep the same time, so having two clocks in the same room, like having two public clocks in the same town, called the authority of both into doubt. Worse, it called the nature of time itself into question.

Mechanical timekeepers, when they appeared in schoolbooks, poetry, and popular literature, usually recalled time on this grand scale—clocks were not portrayed as machines for telling the time of day, but as mechanical embodiments of mortality, symbols of eternity and the brevity of life. The old-fashioned tall case or "Grandfather" clock most often filled this role, creaking its doleful message of regret in countless poems. Longfellow's "The Old Clock on the Stairs," included frequently in children's texts, is typical of the genre:

> Somewhat back from the village street
> Stands the old fashioned County seat.
> Across its ancient porticoe
> Tall poplar-trees their shadows throw;
> And from its station in the hall
> An ancient timepiece says to all—
>   Forever—Never!
>   Never—Forever!
>
> Half-way up the stairs it stands,
> And points and beckons with its hands
> From its case of massive oak
> Like a monk, who under his cloak,
> Crosses himself, and sighs, alas!
> With sorrowful voice to all who pass—
>   Forever—Never!
>   Never—Forever!

Longfellow connected the clock to the New England past he helped create, to the preindustrial rural ideal—the same pastoral vision that appeared on the tablets of Terry's clocks. He also tied it to religion and unknowable mysteries, hence the reference to the monk. But time's face turns grim:

> In that mansion used to be
> Free-hearted Hospitality;
> His great fires up the chimney roared;
> The stranger feasted at his board;
> But, like the skeleton at the feast;

> That warning timepiece never ceased—
> > Forever—Never!
> > Never—Forever!

Longfellow mined this vein for several more verses, his mood growing ever darker till relieved at last by the final lines, which as usual in these poems is resolved by hinting at reunion in eternity.[53]

The somber Grandfather clock spread its pall of gloom on every happy occasion, inflicting itself again and again in children's textbooks. *McGuffey's Third Reader* described how "in the old, old hall the old clock stands, / And round and round move the steady hands." As a little girl stood before it, trying to decipher its meaning, *McGuffey* supplied it: "Tick, tock, tick! Time passes away."[54] When clocks were absent time itself, in the abstract or personified as gray-bearded Father Time, filled the same role. "What is Time?" inquired a poem in one *Reader*. "I asked a dying sinner, ere the tide of life had left his veins: 'Time,' he replied; 'I've lost it! Ah, the treasure!' and he died."[55] Typically baleful if not particularly eloquent, this sinner's last words, or at least his sentiments, might characterize a whole generation's thinking about time. Such gloomy musings filled children's schoolbooks and popular magazines. Why were they so popular?

The idea of time as both a devouring entity and a scarce, precious commodity dated back at least as far as the Renaissance. But these poems also typified the characteristic morbid sentimentality of antebellum culture, the overwrought yearning and nostalgia for better days that Mark Twain later satirized in *Huckleberry Finn's* Emmeline Grangerford—the "Shall I Never See Thee More Alas" school.[56] Exaggerated sentimentality offered a counter to the harsher realities of Jacksonian commerce, and these poems reminded their readers that worldly accomplishments were fleeting. They also tended to reinforce filiopiety—*Grandfather* clocks,

Father Time—the poems connected authority to older male figures, whose experience and memory of better and presumably more virtuous days made their judgments on modern times all the more wise.

But in these poems the paternal clock, paradoxically, also tolls its own death knell. The genre's popularity reflects the decline of traditional paternal authority that characterized the 1830s and 1840s. As the factory system overtook more traditional methods of craft production, it dissolved the paternal relationship between the boss or owner and his employees. Under factory production, instead of working for someone they knew personally, even intimately, employees toiled for a distant stranger whose authority worked its way down a chain of managers and foremen. In these poems, older paternal authority is dead or dying. And the sense of time associated with that authority—old-fashioned clocks, pastoralism, and the cycle of life—threatens to give way before a new order. Death, and happiness under the ultimate paternal authority of God, is the consolation the poems offer for paternalism's decline.[57] With the decline of patriarchal authority came a change in the idea of time and clocks that went with it.

In terms of clock imagery, Longfellow's poem makes an interesting comparison with a common children's lesson, "The Discontented Pendulum." Early one morning, went the tale, an old farm clock suddenly stopped moving. Fed up with mindlessly swinging back and forth in the dark and jealous of the dial's better working conditions, the pendulum had quit. Alarmed, "Mistress Dial" reminded the pendulum that his work, though dreary in the aggregate, amounted to no great effort at each swing, and that all the other parts had their work to do as well, even if theirs did seem more pleasant. Chagrined, the pendulum went back to work. "As if with one consent, the wheels began to turn, the hands began to move," the clock struck, and the farmer got up. The moral of this story was obvious: keep your nose to the grindstone and know your place in the social machinery. Notice that the clock, not the sun or the industrious chirping birds, wakes the farmer up.[58]

The dual role of clocks and time in American culture reveals itself in a comparison of the clocks in the two stories. In Longfellow's poem, the clock insists on the inevitability of decline and mourns the passing of the old order. In the second example, the busy, bustling "Mistress Dial" regulates the new. In Longfellow's poem and the genre it represents, clocks symbolized the natural, cyclical time of the disappearing past. They reminded their readers, through the imagery of the clock, that death was the ultimate, inescapable judgment on human folly. In "The Discontented Pendulum," the clock served as the model for social organization, not because it mimicked nature, or symbolized some gloomy abstraction, but because it reproduced social hierarchies working together harmoniously—the story described the clocklike movements of industry as a positive virtue.

In the story, clocks, not nature or patriarchal authority, regulate life and labor. It was in this sense that Henry Clay enthused: "who has not been delighted with the clock-work movements of a large cotton factory?"[59] Clay's comment, and the story of the disgruntled pendulum, point out the clock's growing role, its increased authority, in ordering daily life and work.

In the 1830s clock time was entering the work place through the factory system. In the first years of factory work textile mills, for example, operated only in sunlight. True, most mills listed precise hours of opening and closing. We have examples of these schedules, and in their typographical finality they make it seem to the modern observer as if the worker's days were strictly counted out in numbers on the clockface. But in fact those hours of work depended on fluctuations in consumer demand and, most notably, available sunlight. Factory schedules invariably changed with the passing seasons, since artificial light was both expensive and dangerous. "The period of labor is not uniform; in some cases from sun to sun," one workman claimed. "It is most common to work as long as they can see."[60] The connection between hours of work and the sun suggested that early factory labor mirrored the cyclical time of farm work to some extent. In many instances, that may have been the case—mill timetables may only present us with an ideal, as fluctuations in available daylight and demand for goods affected the hours of labor.[61]

But as clocks grew more common, artificial light became more practical, and regional trade networks regularized, factories began adopting more rigid hours. The gradual imposition of clock time through the factory system constituted one of the central conflicts of industrialization. Workers struggled hard to preserve some control over the pace and hours of their labor, instead of conceding their schedule to clocks, and the process of resistance and accommodation to factory time has been well documented.[62] But these men and women's objections to the conditions and extent of work—meaning the number of hours demanded, time allotted for meals, work conditions, and so on—should not be automatically mistaken for opposition to regulation by the clock. Resistance to factory time itself is different from the quest for shorter hours, and resistance to clock authority is different from objections to work schedules themselves. There are many examples of workers resisting the boss's demands. Examples of workers debating the nature of time itself are less common, but no less interesting.

With the proliferation of factories and the consolidation of the factory system, clock time became increasingly common in daily labor. *Defining* that time—especially defining time as factory bells—gave owners an edge over their employees. Factory bells could be made to ring at will, and their sounding might easily be delayed to wring an extra half hour's work out of employees who had no reliable way of measuring time themselves. As accustomed paternal relations between labor and management deteriorated, workers began feeling the lack of a reliable standard of time.

In 1832 the New England Association of Farmers, Mechanics and other Workingmen reported with outrage that in Hope Factory, Rhode Island, the time for shutting the factory gates at night was "eight o'clock, by the *factory time,* which is from twenty to twenty-five minutes *behind* the true time," and that in Nashua and Dunstable, New Hampshire, "the *factory time* [was] twenty-five minutes *behind* the *true solar* time." At the Arkwright and Harris Mills in Coventry, Rhode Island, the report continued, "labor ceases at eight o'clock at night, *factory* time." In Pawtucket, the situation was found to be similar, "with the exception that, within a few weeks, public opinion has had the effect to reduce the *factory time* to the *true solar* standard."[63] Obviously, the mill owners were cheating, hiding behind impressive-looking clocks while they tampered with time for their own advantage. Investing the authority of time in clocks gave them an edge, while their workers insisted on connecting time to the impartial sun.

Controlling and defining time translated directly into economic and political power. Workers in Pottsville, Pennsylvania, a mining town, discovered another way clocks might cheat them in an 1843 election dispute. Officially the Pottsville polls closed at seven, but numerous witnesses saw civic-minded voters exercising the franchise till at least eight-twenty. Or so it seemed — "it is well known," claimed the Pottsville *Miner's Journal,* "that we have no exact or certain standard of time in this borough; it is notorious," insisted the editor, that watches and clocks in Pottsville varied by differences "frequently being as much as one hour." An election inspector, using a chronometer regulated in Philadelphia not three days earlier, insisted that the polls had closed at seven. But Philadelphia's time was not Pottsville's, and the losing side mounted a drive to declare the election invalid.[64]

The hearings that followed revealed the multiple sources of time the town used, and the confusion — and political opportunism — that resulted. Several local witnesses deferred to "Heywood and Snyder's Foundry bell" as their source of time. The bell's prominence made it a common point of reference, but at least one local resident admitted that he routinely set his watch "fifteen minutes slower than that bell" because "I was under the impression that the bell was too fast." One watchless voter testified, "I went down to Geisse [a local jeweler] . . . and at his clock it was twenty minutes past eight." "Henry Geisse's clock," he added, "is always ten or fifteen minutes slower than the [foundry] bell." A watchmaker concurred: Heywood and Snyder's bell was "one quarter of an hour faster" than the sundial he used to regulate his shop. The bartender at the local hotel reported yet another time, about 9:00 p.m., on their "best regulated clock," while Nathaniel Mills insisted that "by my clock the election closed at a quarter after seven."[65] Who owned the most reliable watch? Whose clock told the correct time? In Pottsville, with no agreed-upon standard, it was impossible to tell.

These men objected to being victimized by an arbitrary standard of time—a time derived from clocks they couldn't verify or control. Since watches were still too expensive for most people, the sun offered an indisputable standard honored by tradition and religious authority, free from the owner's influence. In 1832, the *Pawtucket Chronicle* had praised a plan to put a tower clock in the Congregational Church, since "all are aware of the vexatious confusion occasioned by the difference of time in the ringing of the factory bells . . . which can only be remedied by creating a clock that will always give *the time of day.*" The clock, paid for by subscription and blessed by location in the church, was to provide an impartial standard based on local sun time, and keep the mill owners from cheating.[66] Tying a clockwork regulator, like the Pawtucket clock, to the rational, impartial, and objective traditions of celestial mechanics—having it show sun time—made it seem benevolent and impartial rather than harmful or oppressive.

And indeed as artifacts, viewed independently of who controlled them, clocks and watches offered a stunning example of human genius and skill. "What a miracle of art," Edward Everett exclaimed to the Massachusetts Charitable Mechanics Association in 1837, "that a man can teach a few wheels, and a piece of elastic steel, to out-calculate himself; to give him a rational answer to one of the most important questions that a being traveling toward eternity can ask!" "What a miracle," Everett repeated, "that a man can put within this little machine a spirit that measures the flight of time with greater accuracy than the unassisted intellect of the profoundest philosophy." Addressing the Massachusetts mechanics on the usefulness of mechanical arts, Everett praised the watch's capacity to serve as a model for rationality and the organization of the mechanic's time. "By means of a watch," Everett enthused, "punctuality in all his duties—which in its perfection, is one of the incommunicable attributes of Deity—is brought within the reach of man."[67] Everett made the watch holy. He reconnected timekeeping machines to heavenly mechanical principles and the rationally ordered society.

Everett's platitudinous address to the Mechanics Association posited a pettifogging deity who listed flawless, clocklike punctuality first among his virtues. It reflected a social need for docile, well-disciplined employees, but it also reflected the further spread of clock time into everyday life. Thanks to cheap, mass-produced timekeepers, the clock time that confronted labor in the work place found its complement in clock-based regulation of the private sphere and the household labor as well.

Home advice manuals of the 1830s focused on systematizing the household economy. Lydia Maria Child's *The American Frugal Housewife* (1835) reminded its readers on page one that "Time is Money," and urged them to "gather up all the fragments of *time* as well as materials."[68] William Alcott's popular handbook *The Young Housekeeper* (1838) began its advice on "having a plan" by recalling the clockwork universe. "Order is heaven's first law," Alcott wrote, "and should be the first law of that which properly managed would, of all places below the sun, most nearly resemble heaven."[69] Alcott recapitulated the clichés of celestial mechanics, but his home resembled heaven because its members, in following heaven's law, had internalized the clock's authority, not natural examples.

Alcott offered detailed advice on waking at regular times. "There is no difficulty in waking at a desired hour of the morning, when one has a strong motive for it," he wrote, but in such cases one sleeps uneasily. Alcott suggested that "there should be some ingenious contrivance, to awaken the housekeeper at first—perhaps an alarm clock . . . till the habit is effectively formed; after which, she will find little difficulty in awakening." Alcott offered the clock as a model of regularity, order, and self-government, and suggested that the machine be fully incorporated into the unconscious. Having internalized clock time, the virtuously punctual could rest easy.[70]

The result, he believed, would secure regular and precise habits at home and in public life. "The hour for breakfast," he asserted, "should also be assigned, and when thus fixed, should not be delayed except on special occasions." It was better in all respects, he continued, "to sit down a few minutes before the time, than even one minute or one half minute later;—better for the body, better for the mind, and better for the whole character." Let these moments be fixed, Alcott continued, "and except in emergencies, as in the arrival of a friend or the occurrence of an accident, let [them] seldom if ever be departed from."[71] Alcott's devotion to punctuality and regularity was so fierce, so religious, that a friend's arrival caused a household emergency.

These domestic management guides point to the continuing evolution of new, clock-based systems of social order. Lydia Maria Child's admonition to "gather up all the fragments of *time* as well as materials" hardly differs from Esther Burr's submission to God in labor, or the almanac's promptings to save time. But Alcott's handy alarm clock allowed a woman to organize the entire household on a machine-inspired model of regularity—the clock governs both the home and the woman who oversees it. Recalling "Mistress Dial" of "The Discontented Pendulum," Alcott's women readers could use the clock to systematize the small hierarchical machinery of the home, in turn preparing their families for regulation in the larger machinery of the public work place. These advice manuals represent, like the factory bells, the growing power of clocks and clock time.

But conflicting beliefs about time's source—natural or mechanical—and its character—cyclical or linear—continued to haunt American culture and politics. The writers of the American Renaissance in particular greatly feared a social time transformed permanently by industry. "For a great many years past," wrote Hawthorne in 1838, "there has been a wood-cut on the cover of the Farmer's Almanac, pretending to be a portrait of Father Time." Hawthorne referred to the traditional depiction of Time as a quasi-deity, a winged figure with an hourglass and scythe. These days, Hawthorne continued, Father Time "has exchanged his hourglass for a gold patent lever watch, which he carries in his vest pocket; and as for his scythe, he has either thrown it aside altogether, or converted its handle into a cane." Hawthorne made Time a prosperous and respectable figure, a man about town strolling the avenues in the latest clothes. If Time's transformation mirrored the farmer's conversion to city life, it also suggested the changed understanding of time. Father Time still cuts down the living in Hawthorne's story, but in 1838 he moves to the bustling pace of city life, with a watch to guide him.[72] Hawthorne offered an ironic commentary on the clock's ascendence in everyday affairs.

Hawthorne's "The Artist of the Beautiful" opens with the artist, Owen Warland, laboring unhappily as a town clockmaker. Endowed with a genius for intricate mechanism, Warland hates the unimaginative work of clockmaking and labors to create beautiful, living "spiritualized mechanisms." "If a family clock was entrusted to him for repair," Hawthorne mused, "one of those tall, ancient clocks that have grown nearly allied to human nature, by measuring out the lifetime of many generations," Warland would install some mechanical whimsy of his own design—hands that spun rapidly about, figures that danced to chimes that rang too many times. "Several freaks of this kind," Hawthorne continued, "quite destroyed [his] credit with that steady and matter of fact class of people who hold that time is not to be trifled with, whether considered as the medium of advancement and prosperity in this world, or preparation for the next."

In the passage above, from 1844, Hawthorne pointed out two contemporary interpretations of the meaning of time and made Warland, the rebellious artist, violate both by taking neither seriously. Warland's spirit, however, is too delicate for sustained rebellion. Repeatedly crushed by the cold-hearted sneering of his former master in the clockmaking trade, a sagacious Yankee interested only in practicality and production, Warland for a time keeps regular hours and even succeeds in repairing and regulating the town clock. "The town in general thanked Owen for the punctuality of dinner time," but Warland's refined creative ambitions return to haunt him. He stops keeping regular hours, he "wastes the daylight" in walks and begins working feverishly at night. He triumphs in the end, making peace with himself and realizing the Enlightenment dream of living mechanisms, but only after completely throwing off the compulsion to regular hours, practical work, and clock time, and freeing himself finally from the need to "progress" in invention. In the end, the real achievement is not Warland's mechanical invention or his concession to social regularity but his own spiritual self-realization.[73]

In "The Devil in the Belfry" (1839), another attack on clock time, Edgar Allan Poe described a small Dutch town, "Vondervotteimittis," literally patterned after a clock. As the story begins, a tall clock tower stands in the center of a valley, and around it "extends a continuous row of sixty little houses. . . . Every house has a small garden before it, with a circular path, a sundial, and twenty-four cabbages." The town looks like a clock, and the clock holds the authority for the organization of both time and space. According to the story, all the inhabitants of the town and even their animals wear watches; they all "wonder what time it is" all the time. At noon every day all action stops as the worthy burghers count along with the strokes of the bell. But one day their orderly lives are disrupted when a small figure appears on the horizon, in the form of a fiddler dressed in black and having "not the remotest idea in the world of such a thing as keeping time." The fiddler ascends to the belfry, begins playing madly, and this time at noon the clock strikes thirteen times, plunging both the townspeople and their machines into chaos.[74]

Poe satirized the authority of clocks, linking them to middle-class complacency and obsession with order. In Vondervotteimittis, political authority, the ordering authority for the town, stems from the master clock. Its satellite clocks in each house, like their owners, are enslaved to it.[75] Poe pursued the connection between political authority and time more subtly in "The Pit and the Pendulum," written in 1843. Here Time's traditional scythe is transformed into a giant clock pendulum with a razor edge. The benign pendulum of Terry's clocks becomes for the narrator, bound to a table in the center of the pit, an instrument of torture in the hands of an invisible political enemy. The time remaining to the victim is tied directly to the swings of the clock pendulum, and not to Time in the abstract.[76]

Poe and Hawthorne, as Romantics, loathed the oppressive power of clock time, its mechanical regulation of life and thought. They expressed resentment and even rage at a familiar bind; unable to support themselves in a trade, writing, that demanded flexibility, freedom, and time for contemplation, they located the source of their oppression in clocks and their unceasing regularity. Both reacted specifically to industrial society and its new controlling device, the mechanical timekeeper. Recognizing time located in mechanical sources as an expression of political, economic, and social power, they interpreted their society's beliefs about time as moral issues involving personal freedom.

Objections to clock time in industry and self-government transcended Hawthorne and Poe's narrow literary circle. The Sabbatarian movement of the 1840s also connected time usage to morality, and recognized tyranny in the ticking clock. The question of how to observe the Sabbath was a major bone of contention for American Protestants throughout the nineteenth century. Mormonism and Seventh-Day Adventism both emerged from Sabbatarian debates over time and how it should be defined. Sabbatarian reformers saw factory work as unpleasant and often exploitive, yet necessary and potentially good. They objected to its apparent Godlessness, and wanted some way of restoring Christian belief to the humming mills. Sabbatarian activists agitated for the suspension of all labor and commerce, by law, on the Sabbath. Sundays, they felt, should properly be spent at home, in thanks to God and quiet meditation on the state of one's soul. Such Sunday laws would bring God back to the industrial week.

"Those who demanded strict Sabbath observance," according to one historian of American religion, "insisted that the principle of a day's rest in seven was planted in the very order of nature and in the original constitution of humanity." Recalling the almanac and the virtue of natural law, the Sabbatarians posited a natural, internal "moral clock" in need of rewinding every seven days.[77] "The Lord's day," wrote one Sabbatarian, "is not merely the day of religious duty and rest, but the restoring, the awakening day—the day of recovery and reformation." Like other Sabbatarians, he imagined a "higher" sense of time that God had built into the machinery of the soul. "If the Sabbath be desecrated," he continued, man's "whole spiritual prosperity and existence are endangered."[78]

Not just spiritual prosperity but financial prosperity as well suffered, for the two were linked. Sabbatarians recognized clock time's ascendency in public life and labor—they knew the factory system was here to stay. But they hoped, by setting Sundays aside from industrial hue and cry, to prevent industrial time from completely overcoming the "natural time" of the soul. Sunday laws, they claimed, would help employers by rejuvenating tired workers and ministering to their moral health, and help laborers by mandating at least one day's rest per week. Keeping the trains from running on Sunday, Sabbatarians hoped, would bring the working classes back into the wholesome sphere of home and church on the holy day, restoring some semblance of God's time to the weekly work schedule. Ironically, William Alcott's example, quoted previously, suggested that home now offered little relief.[79]

But what the Sabbatarians saw as a way to ease the laborer's burden, and regulate his or her moral health, their more extreme opponents rejected as an accommodation with industrial time. Most opponents of Sabbath laws argued that the strict Sabbath observances violated the separation of Church and State, threatened individual liberty, or "paganised" Christianity by imitating Jewish customs. More radical Anti-Sabbatarians loathed Sunday legislation as a dangerous and misleading endorsement of industrial time and the evils it caused.

Convening to protest Sabbath laws in 1848, a group of men and women including William Lloyd Garrison, Francis Jackson, Bronson Alcott, Theodore Parker, and Lucretia Mott attacked

Sabbath legislation as a false friend of the laborer and a tool of oppressive power. Garrison exhorted the working classes to "claim more and take more rest . . . strike at the exercise of that power . . . which is eating up your hard-won substance, and keeping you in shameful vassalage." Do this, he continued, "and you shall redeem all days from servile toil, and enjoy a perpetual Sabbath." Garrison urged industrial laborers to resist the employer's division and control of their time.[80]

Even more explicitly, the Anti-Sabbatarians objected to imposing clock time on Christian practice. One delegate claimed, "Our Sabbatarian friends have a queer way of setting the right and the wrong of actions." According to them, "if you want to know whether it is right to walk out, and take the air on Boston Common, look at the clock! If the finger points to a particular hour, it is very sinful; if it points to another hour, it is perfectly right! If you want to know whether it is right to read a newspaper, or write a letter, to post a letter, or to carry a letter, ask, not whether the act is right or wrong, but ask, what time of day it is, and what day of the week it is! Consult the calender and the clock."[81] Directly attacking clock time and its new role in society, the speaker deplored the clock's place as the arbiter of personal morality and the governor of civil affairs. In his view the once private, unbounded world of ethics and religious feeling now lay exposed, like the narrator in "The Pit and the Pendulum," subject to mechanical regulators unrestrained by religious authority or natural law.

The Sabbath debate revealed the degree to which changing conceptions of time and its source were working to rearrange time in both the public and private spheres. It focused on that new but scarce industrial phenomenon, leisure time—how was it to be spent? Sabbatarians, who accommodated themselves to industry, wanted to shut down industry's most visible symbol, the railroad, every Sunday, to take brief refuge from clock time in the peaceful, "natural" world of the home. Their opponents, who hated industrial excesses and linked them to slavery, hoped by controlling industry to smash the clock and return society, and social time, to its natural origins.

One of the most coherent and sophisticated attempts to comprehend and define social time came from the daughter of a Sabbatarian activist, Catherine Beecher. Regularity and loathing for wasted time formed the core of her 1841 best-seller, *A Treatise on Domestic Economy*. Like Esther Burr, quoted at the beginning of this chapter, Beecher saw work as a duty to God. Her chapter "On Economy of Time and Expenses" reminded her readers, "Christianity teaches that for all the time afforded us, we must give account to God; and that we have no right to waste a single hour." Though she recognized recreation as necessary to health and insisted that play receive its just allotment of time, she urged that women not squander their time in trivial pursuits. Temporal virtue required unstinting watchfulness and discipline, and play remained useful only to the extent that it promoted useful work: "A woman is under obligations to so arrange the hours and pursuits of her family, as to promote systematic and habitual industry; and if, by late breakfasts, irregular hours for meals, and other hinderances of this kind, she interferes with, or refrains from promoting regular industry in others, she is accountable to God for all the waste of time consequent on her negligence."[82] Superficially, Beecher's admonitions resembled the promptings in household manuals like Alcott's and Child's. But Beecher specifically rejected clock time in favor of a virtuous temporal rigor drawn from the Bible.

Women's example for organizing their days, she concluded in a long discourse on the *"right apportionment"* of time, must be the demands in time and property God asked of the ancient Hebrews. "In making this apportionment [of time]," she continued, "we are bound by the same rules as relate to the use of property."

Beecher's connection of time to property acknowledged time's role in labor and commerce, but also recalled women's position in society; Beecher implied that time was women's only real property, since material property typically belonged to men. She wrote to improve the sad state of society as she saw it. Women, as examples of kindly diligence and temporal thrift, were to quietly promote the moral organization of society using the same model of time she suggested for the home.[83]

Beecher emphasized hard work and saving time. But she rooted her conception of time in nature, rather than in clocks — "it thus appears," she wrote on the length of the work day, "that the laws of our political condition, the laws of the natural world, and the constitution of our bodies, alike demand that we rise with the light of day . . . and retire when this light is withdrawn." Beecher revived the cyclical aspect of "natural" time, but she relocated it squarely in the home, not the public work place. In her interpretation of nature's pattern for time usage, women's subservience to man, God, and natural rhythms became a weapon for the ethical reform of society through efficient, diligent, morally justified work in the production of good citizens. The home would serve as the virtuous, rationally ordered nest in which fledgling members of industrial society would learn the right principles for organizing their time.[84]

The immense popularity of Beecher's manual attests to both the applicability of her ideas and the growing division between private and public time. Beecher attempted to reestablish time's authority through religion and natural law, and thereby to make order out of the new ideas of time that prevailed outside the home and now knocked insistently on home's front door. Like "Mistress Dial" of "The Discontented Pendulum," she sought authority to regulate society, but she hoped her elaborate theological justification for time usage would transcend the secular self-interest that characterized other definitions of time, and put time on a less arbitrary basis. She used her emphasis on work to empower women—in the home, women controlled the pace of their labors and decided what tasks to do. In setting the terms of their labor they answered to God, not to factory schedules and clock bells.

Beecher consciously limited her ideas to the home, retreating from the challenges industrial time posed, or rather meeting them only halfway. Her biblically justified time made sense for many women and men, and helped establish a split between home time and public time that persisted long after her book went out of print. Beecher made home the place of "natural" time, morally virtuous hard work, and religious feeling, in opposition to the uniform, standardized public clocks, based on astronomical observation and telegraphed time signals, that were emerging as an alternative for regulating public time even as she wrote.[85]

Seeking, like Beecher, a reliable, uniform, justifiable, and objective standard of time, the 1851 timetable of the Lowell mills specified exactly when the yard gates opened and closed, the times for meals, and exactly how long work began after each bell: "WORK COMMENCES, at Ten minutes after last morning bell, and at Ten minutes after bell which 'rings in' from meals." "In all cases," it continued, "the *first* stroke of the bell is considered as marking the time." The timetable also specified "the Standard time being that of the meridian of Lowell, as shown by the regulator clock of JOSEPH RAYNES, 43 Central Street," so no confusion could result. The Lowell mills allied themselves with astronomy and science to invest their factory bells with an unimpeachable authority.

Based on the meridian of Lowell and the sun, Raynes's clock minimized the possibility of cheating and ensured that in Lowell, noon by the sun and noon by the clock would seem nearly the same. Adopted as the standard of the mills, the clock lent an aura of impartiality and natural law to the work schedule. Earlier, the Lowell owners had relied on paternal supervision to ensure the

TIME TABLE OF THE LOWELL MILLS,

To take effect on and after Oct. 21st, 1851.

The Standard time being that of the meridian of Lowell, as shown by the regulator clock of JOSEPH RAYNES, 43 Central Street

	From 1st to 10th inclusive.				From 11th to 20th inclusive.				From 21st to last day of month.			
	1st Bell	2d Bell	3d Bell	Eve. Bell	1st Bell	2d Bell	3d Bell	Eve. Bell	1st Bell	2d Bell	3d Bell	Eve. Bell.
January,	5.00	6.00	6.50	*7.30	5.00	6 00	6.50	*7.30	5.00	6.00	6.50	*7.30
February,	4.30	5.30	6.40	*7.30	4.30	5.30	6.25	*7.30	4.30	5.30	6.15	*7.30
March,	5.40	6.00		*7.30	5.20	5.40		*7.30	5.05	5.25		6.35
April,	4.45	5.05		6.45	4.30	4.50		6.55	4.30	4.50		7.00
May,	4.30	4.50		7·00	4.30	4.50		7.00	4.30	4.50		7 00
June,	"	"		"	"	"		"	"	"	* "	"
July,	"	"		"	"	"		"	"	"		"
August,	"	"		"	"	"		"	"	"		"
September,	4.40	5.00		6.45	4.50	5.10		6.30	5.00	5.20		*7.30
October,	5.10	5.30		*7.30	5.20	5.40		*7.30	5.35	5.55		*7.30
November,	4.30	5.30	6.10	*7.30	4.30	5.30	6.20	*7.30	5.00	6.00	6.35	*7.30
December,	5.00	6.00	6.45	*7.30	5.00	6.00	6.50	*7.30	5.00	6·00	6.50	*7.30

* Excepting on Saturdays from Sept. 21st to March 20th inclusive, when it is rung at 20 minutes after sunset.

YARD GATES,

Will be opened at ringing of last morning bell, of meal bells, and of evening bells; and kept open Ten minutes.

MILL GATES.

Commence hoisting Mill Gates, Two minutes before commencing work.

WORK COMMENCES,

At Ten minutes after last morning bell, and at Ten minutes after bell which "rings in" from Meals.

BREAKFAST BELLS.

During March "Ring out"........at....7.30 a. m..........."Ring in" at 8:05 a. m.
April 1st to Sept. 20th inclusive.....at....7 00 " " .... " " at 7.35 " "
Sept. 21st to Oct. 31st inclusive.....at....7.30 " " .... " " at 8.05 " "
Remainder of year work commences after Breakfast.

DINNER BELLS.

"Ring out".......... ....,......12.30 p. m.........."Ring in".... 1.05 p. m.

In all cases, the *first* stroke of the bell is considered as marking the time.

1851 timetable of the Lowell Mills. Notice that the hours of operation vary not just with the seasons, but with the days of each month, presumably to follow available sunlight.

(Reproduced courtesy of the Baker Library, Harvard Business School.)

morality of the work that went on there — "The supervisor's mind," it was claimed, "regulates all, his character inspires all; his plans . . . control all."[86] Supervision of the workers' home lives combined with a pastoral setting and the ostensibly temporary nature of the work to "republicanize" the factory town.

But native New England women fled Lowell's exploitative conditions in the 1840s, and Irish immigrants began taking their places at the machines, immigrants with less love for the hard lessons of *McGuffey's Readers*.[87] For the mill owners, astronomically regulated time shored up the Protestant Ethic with "natural law" while putting unruly immigrants on a more fixed schedule. Even better, the new standard time linked the mills and their owners to national scientific movements, especially astronomy, that by the 1850s were being increasingly associated with the economic, moral, and physical progress of the nation.

By synthesizing science, commerce, and mechanical time-keeping, the mill owners "moralized" their regulation of labor time, putting it on a reputedly objective and scientific basis; in this new scheme Raynes's clock, regulated astronomically and hence rational, replaced the human supervisor as the authority governing work. Over the next three decades, astronomy emerged as the regulating principle and source of public time, partially resolving, as Catherine Beecher had for the home, the confusions over time that plagued industrializing America. Rapid communications by telegraph, faster travel by railroad, stage, and steamboat, interstate and interregional commerce all demanded a uniform standard of time. By 1860 astronomical time, telegraphed time signals, and regional standard time had risen to fill the void. In the process, these innovations established new authorities for time, and new models for self-government and social regulation.

II

Celestial Railroad Time

If the years before 1850 saw a growing confusion over time and its source, the next thirty years saw at least a partial solution in the rise of astronomy and astronomically based telegraph time signals from observatories. Astronomical time recapitulated the laws of nature. Legitimized by science, it offered a seemingly impartial and rational standard for regulating commerce and industry. But astronomical time was also closely tied to geographical and commercial expansion, industrial patronage, and personal and corporate gain. The scientists who advanced observatory time signals vied for professional reputation and financial reward. Their work changed the nature of public time, turning a quasi-sacred abstraction into a commodity, a useful tool of industry capable of being owned and sold.

Since David Rittenhouse and well before, Americans had been fascinated with the heavens and tried to interpret their meaning, even applying Newtonian mechanics to political theory. Fittingly, the first public reading of the Declaration of Independence came from the platform of a temporary observatory Rittenhouse had built in the State House yard, to observe the transit of Venus. But astronomy, when linked with timekeeping devices, offered more than a glimpse at the mechanism of the physical world or a model for political organization.

Applied to navigation, astronomical observations facilitated shipping by determining the longitude of prominent shore points and ships at sea. Used in surveying, astronomy helped divide the vast space of the new continent into accurately mapped parcels, rationalizing interstate trade and aiding westward expansion. And applied to industry, as at Lowell, astronomical time eventually became the guiding principle behind uniform, standardized clock time and the operations of the railroads.[1]

Nearly all astronomical measurement requires precise timekeeping, especially determining longitude. Charles Mason and Jeremiah Dixon, the English surveyors sent over in 1763 to settle the boundary dispute between Pennsylvania and Maryland, packed an English astronomical regulator along with them on their muddy colonial mission. They built a temporary field observatory, mounting the precious clock on a log sunk into the ground to minimize vibration. With some preliminary assistance from David Rittenhouse, the intrepid pair then conducted measurements of colonial longitude by comparing local time to Greenwich time as kept by the clock. The legal boundary they established settled questions of political jurisdiction between the two colonies, facilitating further settlement and trade.[2]

The conquest of western space was intimately linked to accurate measurement of space in time. While President, Jefferson oversaw the establishment of the United States Coast Survey in 1807. The Survey sent its minions, with chronometers, to the loneliest, most forbidding places on the continent. They charted the nation's coasts, harbors, and land formations to fix the location of prominent shore points and keep American ships from unexpected encounters with *terra firma*. The Survey also conducted longitudinal studies of points inland for the benefit of mapmakers, explorers, and the settlers who followed. Following custom, the Survey reckoned longitude from established Greenwich meridians.[3]

Toward 1812, as tensions between the United States and England mounted, patriotic Americans began calling for an American National Observatory, reckoning longitude and time from the meridian of Washington. The Observatory, as "an appendage, if not an attribute, of sovereignty," would lessen "dependence" on England and Greenwich while assuring the reliability of American navigational and geographical standards.[4] Though the War of 1812 passed without such an institution, proponents of American distinction in science continued to lobby for a National Observatory—most prominently President John Quincy Adams in his first address to Congress in 1825.

Adams called European observatories "lighthouses in the sky," illuminating man's passage through this world. His phrase connected astronomy to navigation, while his conclusion tied knowledge of the land to national self-knowledge. Without an American Observatory, he claimed, "the earth revolves in perpetual darkness to our unsearching eyes" and we navigate our destiny—not to mention our oceans and muddy frontiers—blindly.[5] Though his address resulted in little action, it anticipated a rising interest in astronomy as a symbol of rational, orderly exploration of the physical world.

Exploiting the new continent's economic potential required accurate information about land formations, resources, watersheds, and indigenous plants and animals. But before settlers or speculators could take advantage of America's earthly bounty, they needed to establish a philosophical or legal order that would prevent the pursuit of happiness from degenerating into chaos. As they moved west, American mapmakers laid out unclaimed land not according to natural features like hills or rivers, but in vast, perfectly square "sections" of 640 acres. These "Cartesian" grids would help ensure a homogeneous, democratic social space for American expansion. The vagueness of laws regarding the settlement and ownership of land had spurred western expansion by giving opportunists room to maneuver. Precision surveying, dependent in part on accurate timekeepers and a standard time of reference, helped resolve legal disputes over title and possession by providing a firm basis, as in the Mason-Dixon Line, for dividing up property.[6] To its most enthusiastic partisans, astronomical time promised to regulate the whole process of exploration, settlement, and industrialization by organizing society on rational principles.

As the conflict between clock time and "natural" time outlined in Chapter One intensified, American passion for scientific astronomy surged. Practices like reading the almanac to puzzle out the meaning of heavenly movements, a common American habit for generations, increasingly came under attack from nascent scientific elites, who deplored its superstitious overtones. Interest in the stars encouraged the foundation of formal, permanent observatories and schools of astronomy. A National Observatory was finally established in 1834, and in that decade Harvard, Yale, the University of North Carolina, and a great many other universities and towns established observatories of their own. Enthusiasm for astronomy grew markedly over the next decades.[7]

In 1843 Adams, now a congressman, marked the founding of the Cincinnati Observatory with an oration explaining astronomy's importance to the new nation. Adams began by recalling the glory of America and its unique political system, which required a rational, enlightened electorate to work properly. He then retraced the history of scientific knowledge and human advancement through a learned review of astronomy and its progress. Claiming an interest in the heavens as a universal human attribute, Adams noted the parallel development of astronomy and astrology with disdain. "We must remember," he asserted, "that of the genuine and the spurious science, of the chaste matron and the painted harlot; the parentage is one and the same." Adams attacked astrology for being, like the harlot, undisciplined and available to anyone. He spurned the common *Farmer's Almanac* and its tables in favor of a "chaste" scientific discipline requiring years of patient courting to master.[8]

Astronomy surpassed all other sciences because "it is from the skies, that the inhabitant of the earth is provided with a measure of time, without which, the mind of man would be unconscious of its own succession of ideas." Without appreciating time, derived from the heavens, man would be "unable to trace the connection between effects and their causes, and incapable of the exercise of reason." Since American government required the exercise of reason from its citizens, and astronomy recapitulated the laws of right reason, Adams reasoned that there was no science "that so urgently needs the protecting, patronizing and encouraging hand of power" as astronomy.[9]

By linking an attack on astrology to his program for publicly supported observatories, Adams aimed to remove control of time from the individual's sundial and almanac, and place it under the "patronizing power" of educated men and the government. In 1843 individual definitions of time, or unregulated time, subverted the order of public life in the same way that "harlots" disordered paternal authority and the home. Longfellow's "The Old Clock on the Stairs" and poems like it had connected time to clocks and clocks to traditional paternal authority, even while mourning that authority's decline. Adams made time female, acknowledging both the decline of the paternal authority embodied in poems like Longfellow's and the popularity of domestic manuals like Catherine

Beecher's. But by pointing out the need to protect time's virtue, Adams empowered a new kind of temporal authority based on scientific education. He took pains to establish the stars as the ultimate timekeepers, but insisted on the need for trained (male) observers to interpret them. Scientific observation of the heavens, and astronomical time, would regulate industrial culture in place of traditional paternal authority.[10]

What made scientific public time such an urgent need? Part of the answer lies in the confusion over time cited in Chapter One. But the new relationships between time and space developed by American commercial expansion probably played a larger role. Travel in the early nation was arduous and slow. In 1790 it took five full days for news to travel from Philadelphia to New York, and fifteen days for the same tired news to reach Boston.[11] But the surveying techniques that made American wilderness accessible furthered interstate commerce by helping in the design of roads and canals. Canals, by linking distant regions, promoted national cohesiveness — Robert Fulton, engineer and canal builder, concluded in 1807 that "when the United States shall be bound together by canals," with their corresponding market relations, "it will be no more possible to split them . . . than it is now possible for the government of England to divide and form into seven kingdoms."[12] Road building surged as well; by one estimate, the number of postal routes increased over eighty times between 1804 and 1834. By 1817 Philadelphia newspapers made the trip to Boston in under ten days, half the time of 1790, and by 1840 they traveled the same route in less than five. A journey from New York to Cincinnati that took three weeks in 1800 took just a little over seven days by 1830.[13]

But close and rapid connection brought problems as well; it demanded an unprecedented degree of organization across ever-increasing geographical space, and the market economy, wherever it spread, profoundly altered social relations. No technological innovation reflected both the promise and the problems of interregional commerce better than the railroad.

Early railroad builders all grappled with a basic functional dilemma. Should railroads be built for speed, simple transport from point A to point B as quickly as possible? Or was the iron track better suited to meandering along between each little market town? Although expected to regularize commerce and run on schedule, early railroads often disappointed their backers. Between confusion about their purpose and frequent mechanical failures, the first railroads were hardly paragons of regularity—the postmaster general concluded in 1835 that the railroads "cannot be relied upon with that degree of certainty, which is all important in the transmission of the mail."[14] In the beginning the quality of railroad service, clattering along at ten miles an hour, mattered less than the fact that it ran at all.

But the railroad's failure to reach the ideal detracted little from its revolutionary impact. Between 1840 and 1860 total American railroad mileage increased more than ten times, while average speeds doubled.[15] Its tracks carved up the land, its smoke and noise filled the air; its speed and range extended the businessman's realm of action. The ambivalent cultural response to the problem of the railroads has been well documented, and needs no detailed review here. But one good example of a different understanding of the new machine—and how unsettling to the experience of time it could be—appears in Hawthorne's 1843 story, "The Celestial Railroad."

In Hawthorne's nightmare revision of Bunyan's *Pilgrim's Progress*, the railroad compresses the "natural" moral journey through time established in Bunyan's classic, the patient and brave endurance of life's inevitable troubles and temptations. The moral shortcuts offered by Mr. Smooth-it-away, Director of the Celestial Railroad, completely overturn the traditional relations between good and evil—Apollyon, Christian's old enemy in Bunyan's tale, now drives the iron horse to heaven, racing past the few humble, old-fashioned pilgrims who still walk the traditional slow path to salvation. Mr. Smooth-it-away's cars dash easily through tribulations that sorely tested Pilgrim's faith, allowing Hawthorne to reduce Bunyan's lengthy allegory to a few pages. Although the dreaming narrator, a passenger on the celestial line, fears for his soul at points, he takes comfort in the dazzling efficiency of technology; the holy parchments that guided Christian are now pasteboard tickets, "much more convenient and useful along the road," and the Valley of the Shadow of Death now blazes with gas lamps. But the railroad's promises turn out to be empty as its vaunted speed, out of control, resolves finally into a headlong rush toward an industrial hell. Two humble pilgrims, who insisted on making the journey on foot, enter joyfully into heaven just as the trusting narrator is swept onto a sinister, smoking steam-ferry and certain damnation.[16]

Hawthorne condemned not just the railroad, but industrial society and the factory system as well. The moral landscape of Hawthorne's story, leveled to build the railway roadbed, offers few reliable signposts. Christian's encounters through life's slow journey—their homely familiarity, their inevitability—collapse before a whole new set of mechanical temptations to faith, the most tempting of all being the convenience of speed. Hawthorne's railroad of 1843 stands for pure reckless speed, rather than punctuality or scheduling, which imply speed under control. The world of the story is a world of revolutionary and dangerous industrial forces running unchecked, not power systematically and willfully imposed.

Published in the same year as "The Celestial Railroad," Adams's Cincinnati speech addressed similar fears. Hoping to check the locomotive rush of invention and regulate it with reason, Adams promoted a new temporal authority for daily life, an authority that used astronomical time as its mapping and regulating principle. Edward Everett, the Massachusetts senator and sometime editor of the *North American Review*, gave a similar address dedicating Albany's Dudley Observatory in 1856. Everett, one of the most popular orators of his day, defended the high cost of stargazing as Adams did. In an age of ever more common and rapid railroad travel, Everett reminded his audience, clocks regulated to local time were useless. When carried east or west they "will keep home time alone, like the fond traveler who leaves his heart behind him."[17] In the thirteen years between the two speeches American railroads had experienced continued growth. But as Everett's comment pointed out, when individual railroads grew and expanded their areas of operation they ran into more and more difficulties with the lack of standardized time.

Running north to south caused relatively little trouble, since the differences in time between individual towns of the same approximate longitude was slight. Moving east to west, on the other hand, highlighted differences of meridian and thus of time. A traveler on a westbound train, setting his watch at departure, might find after less than half an hour's travel that his watch and the local time no longer agreed.[18] To make matters worse, individual railroad and steamship lines each ran by their own standards of time—usually the time of the city the line originated in. When two lines met, or shared a track, or terminated at a steamship landing, it threw differences in timekeeping into high relief. Could a railroad passenger arriving in Pittsburgh at 1:30 p.m. New York local time catch a connecting train leaving Pittsburgh at 1:30 p.m. Pittsburgh time? Confusing at the least, the situation was partially resolved by the appearance, in the late 1840s, of monthly traveler's guides, bound collections of railroad and steamship schedules.[19] But these schedules raised more troubling questions. What time prevailed when two or more standards conflicted? Did the railroad and steamship set the time, or did the local sun?

Such confusion, like the conflicts over time's authority discussed in Chapter One, demanded uniform standards of time usable by all. Like Adams, Everett insisted on a basis in astronomy. "Our artificial timekeepers," he noted, "are but a transcript, so to say, of the celestial motions, and would be of no value without the means of regulating them by observation." Everett stressed how completely, by 1856, "the daily business of life is affected and controlled by the heavenly bodies. It is they and not our main-springs, expansion balances, and our compensation pendulums, which give us our time."[20] Like Adams, Everett advanced celestial mechanics as an ordering principle. And like Adams, Everett favored taking control of time out of the individual's hands and placing it in the care of scientists and learned men. But Everett's speech, with its references to clock time, reflected the continuing spread of Eli Terry's cheap clocks. And his example of traveler's time recalled both the unprecedented growth of American railroads and the troubling profusion of authorities for time.

Everett's awareness of time, astronomy, and railroad travel came from personal experience—he had served as president of Harvard around the time that William Cranch Bond, astronomer and clockmaker of Boston, established the first regional time standards for the New England railroads. Hoping to improve punctuality and safety while ending the confusion over time, in the 1840s a number of railroads issued detailed regulations specifying the time to be used on their lines. The New England roads turned to William C. Bond for their standard.

Bond, a highly successful maker of fine marine chronometers, had assumed charge of the skeletal Harvard Observatory in 1839. There he soon managed to acquire one of the largest telescopes in the world. He also took part in longitude measurements for the United States Coast Survey, where he used a marvelous new invention, the telegraph. American astronomers working, like Bond, for the Coast Survey, had recognized early on that they could use the telegraph to send time signals between their observatories. The time signal made it possible to determine the longitude of one point relative to another — the longitude of Cambridge, Massachusetts, for example, relative to Washington — with far more accuracy than before. From his experience with the Survey Bond invented the drum chronograph, a new, telegraphic method of recording time in the "American Method" of astronomical observations that netted him a medal at London's 1851 Crystal Palace Exhibition.[21]

Railroads, canals, roads, and steamships increased the speed of travel and the flow of goods between regions. The telegraph, perhaps the most marvelous invention of all, bound that network of transportation and trade together. Thanks to the telegraph, news of business transactions and market prices now traveled with lightning speed; so too events in distant places gained a new and pressing relevance for those waiting at the end of the wire. Oliver Wendell Holmes found in the telegraph a "network of iron nerves which flash sensation and volition backward and forward to and from towns and provinces as if they were organs and limbs of a single living body."[22] But telegraphic connection highlighted differences in time even more dramatically. What time was it in Boston when banks closed in Philadelphia? Astronomers like William Bond and his son George began using telegraphed time signals to augment the synchronicity Holmes applauded, and to organize some of the temporal chaos by establishing uniform regional public times in place of local variation.

Bond's eminence as an astronomer, clockmaker, and telegraphic innovator led the New England Association of Railroad Superintendents, meeting in Boston in hopes of consolidating their running times, to recommend in 1849 that "all the railroad companies in New England [adopt] a time two minutes after the true time of Boston as given by William Bond & Son, No. 26 Congress St., Boston." Just why they chose a time two minutes after Boston time remains a mystery. But the Superintendents' determination to formalize timekeeping seems unmistakable. They further recommended that "all station clocks, conductor's watches, and all time tables and trains should be regulated accordingly" to the Bonds' clock. By 1851, railroad officials received daily time signals,

William C. Bond's apparatus for "The American Method"
of determining longitude. Bond's work stood at the intersection of
commercial expansion, rational government, and national identity.
He won a prize for his inventions at the 1851 London Crystal
Palace Exhibition. (National Museum of American
History, Smithsonian Institution.)

by telegraph, from the Bond shop and later the Harvard College Observatory itself. The relationship between the Bonds, the university, and the railroads in effect established the first regional time zone in the United States.[23]

The railroads fulfilled Adams's vision of American commercial life ordered by scientific principles. They drew their time from the Harvard Observatory, standardizing their operations and coordinating their movements through a uniform standard of time. But in the process they attacked the rational appreciation of the heavens Adams treasured. Everett had pointed out that clocks carried east or west soon lost all meaning. It was to synchronize clocks that the railroads adopted the Bonds' standard. By tying their time to the standard *clocks*, the railroads—the preeminent "machine in the garden"—in effect severed the connection between time and nature. Their resolution made the railroad's clock the source of time.

"I watch the passage of the morning cars," wrote Thoreau in *Walden*, "with the same feeling that I do the rising of the sun, which is hardly more regular." As Leo Marx notes, Thoreau understood the railroad through the master metaphor of the clock; in his case, the railroad clock that increasingly ordered life. The clock, the railroad, and their effects on time pervade *Walden* as they did Concord.[24] "The startings and arrivals of the cars," Thoreau claimed, "are now the epochs of the village day. They come and go with such regularity and precision, and their whistles can be heard so far, that the farmers set their clocks by them, and thus one well-conducted institution regulates a whole country." Writing in Concord in the early 1850s, Thoreau described the results of the Bonds' cooperation with the New England railroads.

Thoreau's railroad, unlike Hawthorne's, controlled and regulated its speed and thereby regulated society. "Have not men improved somewhat in punctuality since the railroad was invented?" he continued. "Do they not talk and think faster in the depot than they did in the stage office? There is something electrifying in the atmosphere of the former place."[25] It was the railroad that worked this change in the way people understood and used time. Thoreau often wrote about time in terms of natural law and the conventions of pastoralism—earlier in *Walden* he describes hearing the nearby village tower clocks, "a sweet, and, as it were, natural melody, worth importing into the wilderness." The clocks merge with the pastoral scene, and even the whippoorwills follow this natural time. "They would begin to sing," he slyly observed, "almost with the precision of a clock" each evening at seven-thirty.[26] Thoreau acknowledged the railroad's spiritual origins in Rittenhouse's orrery and Newtonian mechanics—he described "the engine with its train of cars moving off with *planetary* motion" (my italics). But later in the same sentence he corrected himself. The railroad, he observed, followed the course into the future like an arrow—"or, rather, like a comet."

Thoreau went on to compare the trains to bolts shot from some archer's bow—the railroad, he was claiming, embodied the cyclical movements of nature where nature was regarded as a mechanism. A railroad engine's gears and wheels moved in predictable and regular cycles, like the Newtonian universe. But the train of cars had broken the cycle of seasonal time and headed off into the future; it moved, like an arrow, with frightening energy and speed toward an uncertain and potentially disastrous future. He ended by grimly concluding, "we have constructed a fate, an *Atropos*, that never turns aside. (Let that be the name of your engine.)"[27]

Thoreau's famous description sums up the changing American experience of time: an increase in punctuality, surely, but more subtly a shift in the nature and source of time itself, with corresponding changes in work and life. His deliberate and sly comparisons between the railroad, nature, and the regularity of men and the heavens reflected the new public face time began assuming in the 1850s. Astronomical time tied the railroads to natural law and rationalized their operations. But its use in daily life, on the railroads or in the Lowell mills, invested more and more power in what Everett had called a mere "transcript" of heavenly motions—the clock.

The following running regulations, issued by the Camden and Amboy Railroad in 1853, dramatize the effects of the new thinking about time and timekeepers. "Each Conductor, Engine Driver, Switch Tender and Bridge Tender," the rules began, "will be supplied with a good watch by the company." Previously the company had been mostly indifferent about whether or not its employees carried watches. Now they even specified the source of time for each watch, and how it would be secured: "the CONDUCTOR will call for and receive two Watches at the office, preceding each departure—compare them with the Standard Clock, and be sure they agree before leaving the room." The conductor was then to hand one watch to the engine driver, receive it back again at the end of the route, and finally return both watches to the office. Along the way, regulations instructed the conductor to "compare his time with the Clocks at the Stations he stops at, and notify the Agents of any discrepancy."[28]

The regulations make perfect sense. How could a railroad be run without such precision? But by using precise and accurate time as a justification, regulations like these construct an elaborate hierarchy and daily ritual for conferring power, in which authority comes through carrying or reproducing the master clock. The workers' jobs are controlled by the clock, and they in turn spread the authority of the master clock to points along the line. Schoolbooks earlier connected moral virtue to the wise use of time as seen in nature's example. Under these rules wisdom resides in the regulating clock, and virtue lies in the good worker's obedience to it.[29]

═══

The public debate over time's source and its meaning might help explain the peculiar analogy Herman Melville chose for *Pierre*, in his pivotal chapter on moral authority, "The Journey and the Pamphlet." The pamphlet in question, "Chronometricals and Horologicals," by "Plotinius Plinlimmon," compares God's moral precepts to Greenwich time, the navigator's standard, and Christ to a fine chronometer, regulated, like a ship's clock, to Greenwich time. Men and women on earth, in Plinlimmon's analogy, correspond to ordinary watches, each regulated to the more precise standard occasionally but inclined to run fast or slow and drift from Greenwich. Individual consciences, like individual watches, vary from the heavenly standard of morality as watches drift away from Greenwich time.

Moreover, Plinlimmon maintains, the Greenwich standard itself, like Christian morality, stops making sense in other countries, other cultures—keeping English time in China denies the "rightness," the native logic, of China's own standards of time and of personal conduct. "In an artificial world like ours," Plinlimmon claims, "the soul of man is further removed from God and the Heavenly Truth, than the chronometer carried to China, is from Greenwich." For Plinlimmon, "As China watches are right as to China, so too the Greenwich chronometers must be wrong as to China," and so "no single earthly moral position may predominate over another."[30] Trying to keep to celestial standards of conduct in a world so distant from God mirrors the foolishness of keeping Greenwich time in China, and so Plinlimmon advocates an intellectually lame "virtuous expediency," a compromise between God's standards and local realities.

Melville's experiences as a sailor obviously familiarized him with Greenwich time, and the "Chronometricals and Horologicals" conceit seems like a somewhat playful reflection of his years before the mast. But treating the chapter this way ignores the social context he drew on when he chose such a peculiar analogy. On one level, Melville's chapter discussed morality, but at a second level, Melville questioned time's role in daily life. By comparing morality to standard time, the chapter reiterated the connection between metaphors for time, social regulation, and personal conduct. Writing in 1851, Melville satirized redefinitions of time like those engineered by the New England railroads and William Bond.

Even more, Melville probed the connection between time and political authority. The metaphor relates standards of time to personal conduct and morality—just as we use Greenwich time to navigate our way through the stormy seas (and to rationalize land divisions), so we need a reliable standard to use in dividing our time during life's moral squalls. A trip to London in 1849–50 must have acquainted Melville with the recent introduction of Greenwich time as the standard for all of England. Only one year before, most British railroads had set their clocks to Greenwich, in defiance of local time and tradition. The controversial move brought a host of objections, still lingering during Melville's visit. Some condemned "railway-time aggression," others complained that placing Greenwich time over the local standard was "usurping the power of the Allmighty."[31] The authority for time, be it Greenwich Observatory, the railroad, or God, informed the principles, moral, economic, or otherwise, that governed daily life.

Greenwich time symbolized England's maritime dominance— when American sailors consulted their almanacs for information about celestial movements or tides, they made their calculations against the English time. As Melville worked on *Pierre*, patriotic stargazers were lobbying Congress for an *American* prime meridian based on Washington instead of Greenwich. After much agitation from 1846 to 1851, Congress authorized the publication of an American nautical almanac to be reckoned from some spot on American soil.[32] The *American Ephemeris and Nautical Almanac* would help establish the scientific credibility of the United States and formalize an *American* standard of time for commerce and surveying. Though its supporters claimed the American Prime Meridian contributed to the national glory, the plan provoked "great excitement" and widespread hostility from businessmen and shipping interests, who objected to the change from established practice.[33]

Melville's richly ambiguous satire pointed out the political and ethical coercion inherent in standardized, government regulated time in a society, like America in the 1850s, that linked strict attention to time with moral virtue. If a man "seek to regulate his daily conduct" by Christ's chronometrical soul, according to Plinlimmon, "he will but array all men's earthly timekeepers [souls] against him, and thereby work himself woe and death."[34] The way people used time was central to the way society judged them. Deviation from the standards of public time, like deviation from established morality, marked the dissenter as a danger. Through Plinlimmon, Melville satirized the idea of standard public time based on machines.

Plinlimmon's "virtuous expediency" paralleled the system of astronomically regulated, telegraphed time signals recommended by Adams and Everett and adopted by the New England railroads. Virtuous because regulated astronomically, the railroads' time was expedient in that it ignored the "truth" of local time solely to expedite commerce.[35] But while Plinlimmon espoused flexibility, within limits, in moral and temporal affairs, the railroads rewrote the terms of his argument; the railroads were becoming the chronometrical standard *and* managing to impose uniformity. Made specious by changing times, Plinlimmon's philosophy answered nothing. As Thoreau suggested and the Camden and Amboy running regulations pointed out, the railroads' "virtuous expediency" in timekeeping imposed a new, clock-derived logic on the ordering of human affairs.[36]

As Melville published *Pierre* yet another writer addressed the standardization of time, linking it to the health and progress of the nation. "The railroad system of the United States," railway editor Henry Poor claimed, "is destined to be one of the most potent physical instruments in the onward course of the nation. Any improvement that will give to this mighty engine of modern civilization increased safety, efficiency, and order, should be promptly adopted." Time was the key innovation in his opinion, for the railroad demanded close attention to time from its employees—"it is necessary that they should regulate all their movements in time."

A single clock, linked to others, would oversee both worker and trains' movements. "The magnetic telegraph," Poor continued, resorting to organic metaphors that linked man and machine, "the nerve to this vast muscle," would serve to synchronize all clocks. "And before many years," he concluded, "the beats of the great regulating clock in the heart of New York will pulsate with the vibrations of its pendulum on the clock escapements from Halifax to New Orleans."[37]

Poor's vision, of unified national time regulating industry, commerce, and the individual, was premature. National standard time was more than sixty years away, and the railroads deferred any unified action until 1883. But the growth of the railroad and telegraph network—and the public's increased participation in its benefits—dramatized the lack of a nationally uniform, easily distributed time. For example, in 1869, as the transcontinental railroad neared completion at Promontory, Utah, its backers began

planning a national celebration of their achievement. They prepared a special sledgehammer, connected by wire to a nearby
telegraph transmitter. Swung by the soft-handed Leland Stanford,
the sledgehammer would electrically signal its contact with the
last "Golden Spike," and the railroad's completion, to an anxiously
waiting nation. Transmitting the exact moment of completion by
telegraph, they reasoned, would join the entire country, urban
and rural, great and small, in celebration of the iron link between
east and west, while dramatizing the new interconnectedness technology promoted. It would be perhaps the first experience of true
simultaneity in American history.

Fate, and Stanford's unfamiliarity with sledgehammers, intervened: he missed the spike completely. The wired sledge, striking wood, transmitted only a dull thud. A more experienced official
drove the last spike home with a regular sledge as a telegrapher
cabled "done" to waiting stations east and west. But even then
confusion reigned: what time was the spike actually driven? So
many places with so many watches, and no official standard of
time—at Promontory the telegraph clicked around 12:45 p.m., in
Virginia City at 12:30; San Francisco had "precisely" either 11:46
or 11:44:37, depending on which paper one read, while in Washington, 2:47 seemed approximately right.[38] Railroad and telegraph
officials could imagine a simultaneous, synchronized "media event"
taking place across the continent and the world. But the nation
lacked both the mechanical and intellectual precedents necessary
to accomplish it. Telegraph transmitting equipment was still too
crude, the Babel of local times too chaotic.

In the years between 1850 and 1880, following William
Bond's example, other American observatories began building
standard time services: the Dudley Observatory in Albany, New
York, in 1856, the Cincinnati Observatory in 1860, the Naval
Observatory in 1865, and Chicago's Dearborn Observatory
sometime before 1871. Each standardized and regulated, to some

degree, the time of the immediate area and in most cases the railroads nearby.[39] Samuel P. Langley's career in Allegheny, Pennsylvania, dramatizes the observatories' effect on public time after the Civil War.

In Pittsburgh, a series of lectures on the celestial science in 1859 had lead a group of prominent Pittsburgh citizens, including railroad magnate William Thaw, to form the Allegheny Telescope Association in hopes of bringing astronomy to Pittsburgh. Although it acquired an excellent telescope and a serviceable building, the Association failed as popular enthusiasm for astronomy ebbed. When the original astronomer quit in disgust, the growing industrial town was left with a practically useless instrument. After ceding their creation to the Western University of Pennsylvania, the struggling Observatory's trustees began searching for a new director to reorganize its operations.

Langley, the future Secretary of the Smithsonian Institution, came highly recommended from Harvard College Observatory, where he had served as an assistant. Hired in 1867, he spent his first months improving Allegheny's spartan facilities, soon managing to ally himself with both William Thaw and the Coast Survey. He persuaded Western Union to connect the Observatory with the city, entering into a series of observations designed to precisely fix its longitude. The Observatory was then to be used by the Coast Survey to determine the location of other inland cities.[40] Most of the money for this work initially came from Thaw, himself an amateur astronomer. Only Thaw's flowing cash and social influence kept the Observatory afloat.

The struggle for funds plagued scientists everywhere, but American citizens disdained supposedly "impractical" research with special vigor. An ambitious man, Langley recognized that he would never make a name for himself in the combative world of American science if he spent all his time scrounging for dollars or performing tedious longitude calculations for the Coast Survey.

He needed to set his new Observatory on a financially secure basis, independent of public and private largess. It would not be easy. What need had Pittsburgh, a booming industrial city with relatively few cultural pretensions, for a telescope and a professional stargazer?

Langley's clever innovation lay in his decision to *sell* observatory time signals, by telegraphic transmissions, to Pittsburgh's industries. Drawing on his experiences with William Bond, Langley first selected a reliable system for telegraphic time distribution developed in England. In 1869, he issued a pamphlet proposing that Allegheny supply standard time to the mighty Pennsylvania Railroad. Thaw's business association with A. J. Cassatt, president of the Pennsylvania, got the proposal a more than fair hearing. In early 1871 Cassatt issued orders making Allegheny Observatory time the standard for the Pennsylvania Railroad. Langley converted Allegheny meridian time to the longitude of Philadelphia, and sent it to the railroad in return for $1,000 a year.[41]

Time for sale: Samuel Langley, probably in the years he was working on time transmission at Allegheny. (National Museum of American History, Smithsonian Institution.)

Langley also convinced a number of local jewelers to buy time signals at $500 per annum—a practice that jewelers in most other cities eventually adopted. In nineteenth-century America, jewelers were the primary suppliers of high-quality watches and clocks. Seeing a prominent clock labeled "standard time" in the shop window suggested that this jeweler carried accurate goods. It also tended to make the jeweler's store a source of accurate time for passers-by. William Bond had presented visitors to his watch and clock shop with elaborate clocks that proudly showed "Bond's Standard Time." Langley must have made the practice especially attractive to Pittsburgh jewelers.

He also began explaining the idea to the people of Pittsburgh. A series of articles in local newspapers laid out his views on this "startling and novel" proposition.[42] Langley reminded readers that "from the Atlantic to the Mississippi, every city and town is keeping a time of its own." He proposed to "do away with the present vexatious changes of time" in railway travel by overturning the "custom" maintaining local times. Such times, he claimed, were "founded in no natural necessity" and merely a legacy of the past. Langley recalled the difference between mean and sun time. "The time kept by our clocks and watches," he insisted correctly, "is a much more artificial thing than is generally supposed. . . . *No natural phenomenon designates the hour of noon.*" Local noon was a "fiction" by the clock. And since "there is no natural necessity for repeating this fiction . . . there is consequently no reason why Philadelphia and Pittsburgh clocks should not be set to agree with each other." Langley moved to replace natural phenomena with scientific and commercial rationality.

Recognizing the railroad as the key to Pittsburgh's industrial economy, Langley also recognized its role as titular public timekeeper. But he needed to persuade the railroads, and the public, that his standard time offered more than the haphazard system then in use. John Quincy Adams had connected the stars to rationality and virtue, but in his advocacy of Allegheny Observatory time Langley promoted accuracy and precision above all. His clock, he claimed, having been "rendered practically perfect by the hourly comparison with the heavenly bodies . . . will be kept within a second of absolute truth" every year.[43] "Absolute truth," in this case, meant truth as to the meridian of Allegheny; the system's accuracy and universal applicability assured its value to society, its "moral truth," as well.

Langley's novel idea made time a commodity. Local time, as a "fiction," paled besides Allegheny's new and improved version — Langley called it "a relic of antiquity" like local coinage or local weights and measures, which "the progress of centralization, and the interchange of commerce and travel" outmoded. Traditional local time expressed a relation between the individual and the place he or she lived; time of day depended on where you were at a particular instant. Its authority rested entirely on inherited tradition and customs of the past. It became "fictional" when two or more times disagreed, when two possible "presents" confronted each other. Standardized time, on the other hand, expressed a relation between the individual and an entire region — it acknowledged the links between places forged in commercial expansion. Railroad men like Cassatt and Thaw traveled — they administered business empires covering vast stretches of territory, each part dependent to a considerable extent on movements along another. Scientific hourly signals established a more comprehensive "present" equally useful in all places.

By emphasizing the service's scientific accuracy, Langley offered Cassatt and Thaw a rational aid to the problems of controlling such a vast empire. It aided further commercial and geographic expansion. Allegheny Observatory time eliminated the paradox that worried Melville's Plinlimmon; Langley's service derived its authority from his accuracy and its "nonfiction" from its supposedly universal applicability. Standard observatory time made it possible for a centralized railroad organization to oversee and synchronize its many moving parts, human and otherwise. It became the standard of conduct for railroad employees.

Like standard money, this new time also lent itself to circulation. Easily transmitted by telegraph, it could be bought and used by anyone whose line of work demanded it. In Philadelphia Henry Bentley, an enterprising businessman, connected his Philadelphia Local Telegraph Company to the clock the Pennsylvania Railroad used to receive Langley's time signals. He then offered Langley's astronomical time to other local businesses as "Philadelphia Standard Time." Through his central office Bentley provided a synchronized "telegraphic exchange." Subscribers could conduct transactions, send or receive news, and keep abreast of commodities prices by wire.[44] Langley compared standard time to standardized denominations of currency—an apt comparison, since telegraphed standard time facilitated the same exchange of goods and information that money encouraged. Standard time, like standard money, was a universal solvent dissolving the glue of local tradition and custom.

Langley's innovation made economic and scientific sense, and other observatories quickly adopted their own versions of the Allegheny system. Over the next ten years, Langley and his fellow astronomers began organizing, promoting, and selling first time services and later national standard time obtained from observatories. Each assumed that the superior accuracy and convenience of their services doomed "free" local time to extinction.[45]

Others, not surprisingly, held more tightly to local time. Earlier in 1869 Charles F. Dowd, an amateur scientist and principal of Temple Grove Ladies' Seminary in Saratoga Springs, New York, had begun advancing the first plan for uniform national time, one involving four standard time zones remarkably similar to those in use today. His scheme asked all railroad lines to adopt the meridian time of Washington, D.C., and the Naval Observatory. This "railroad time" would be distinct from local time, with the difference to be posted at each station. The energetic Dowd calculated a set of "plus or minus indexes" for virtually every train station in the country—roughly 8,000 separate calculations—and mailed off the resulting large pamphlet to managers of the leading roads.[46]

Dowd correctly recognized two distinct kinds of time, one rooted in nature and tied to place, the other springing from telegraphy and rapid travel between distinct places, and little affected by natural imperatives. He kept them separate. Emphasizing the autonomy of small towns, Dowd's plan invested the authority for time and timekeeping in tradition and local custom rather than in scientific measurement.

Although his plan received favorable notices from a number of railroad managers, no action resulted. The rapacious train lines of the 1870s warred more often than they cooperated, and the scope of Dowd's plan required more united action than normal. Posting "plus and minus indexes" posed an expensive and troublesome solution to a relatively minor problem.[47] More importantly, his emphasis on retaining local times jarred with both regional time systems like Allegheny's *and* the railroad's apparent tendency to regulate the public time. Dowd continued to lobby railroad managers over the next ten years, but as an outsider in both the railroad and scientific communities he remained a marginal figure.[48]

By the mid-1870s, most railroads operated within a specific region, and like the Pennsylvania Railroad used a standard of time derived from a nearby observatory or city. In effect, they had developed a system of regional standard times. Smaller lines adopted the time of their larger neighbors, and the relatively few frequent travelers probably either stayed within one region during each day's travel or else knew enough about time differences to adjust for them.[49]

ADIRONDACK RAILWAY.
STATIONS.

^{0}Saratoga^{-12}
^{0}Greenfield^{-13}
^{0}King's^{-13}
^{0}South Corinth^{-13}
^{0}Jessup's Landing^{-13}
^{0}Hadley^{-13}
^{0}Stony Creek^{-12}
^{0}Thurman^{-13}
^{0}The Glen^{-13}
^{0}Johnsburg^{-12}
^{0}North Creek^{-12}
^{0}Fourteenth^{-12}

ALABAMA AND CHATTANOOGA RAILWAY.
STATIONS.

^{1}Chattanooga^{-21}
^{1}Wauhatchie^{-21}
^{1}Morganville^{-21}
^{1}Trenton^{-20}
^{1}Dademont^{-20}
^{1}Rising Fawn^{-20}
^{1}Cloverdale^{-20}
^{1}Eureka Coal Mines^{-20}
^{1}Valley Head^{-22}
^{1}Hollomen's^{-25}
^{1}Brandon's^{-25}
^{1}Portersville^{-25}
^{1}Collinsville^{-25}
^{1}Greenwood^{-25}

ALBANY AND SUSQUEHANNA RAILROAD.
STATIONS.

^{0}Albany^{-12}
^{0}Adamsville^{-13}
^{0}Slingerland^{-12}
^{0}New Scotland^{-12}
^{0}Guilderland^{-12}
^{0}Knowersville^{-12}
^{0}Duanesburg^{-12}
^{0}Quaker Street^{-11}
^{0}Esperance^{-11}
^{0}Schoharie Junction^{-11}
^{0}Schoharie^{-11}
^{0}Middleburg^{-11}
^{0}Howe's Cave^{-10}
^{0}Cobleskill^{-10}
^{0}Sharon Springs^{-10}
^{0}Cherry Valley^{-8}

Albany and Susquehanna Railroad.
(Continued.)
STATIONS.

^{0}Richmondville^{-10}
^{0}Carryville^{-9}
^{0}East Worcester^{-9}
^{0}Worcester^{-9}
^{0}Schenevus^{-9}
^{0}Chaseville^{-9}
^{0}Maryland^{-8}
^{0}Colliers^{-8}
^{0}Cooperstown^{-8}
^{0}Emmons^{-8}
^{0}Oneonta^{-8}
^{0}Otego^{-7}
^{0}Wells Bridge^{-7}
^{0}Unadilla^{-7}
^{0}Sidney^{-6}
^{0}Bainbridge^{-6}
^{0}Afton^{-6}
^{0}Harpersville^{-6}
^{0}Tunnel^{-5}
^{0}Osborn Hollow^{-5}
^{0}Port Crane^{-5}
^{0}Binghamton^{-4}

ALEXANDRIA, LOUDOUN AND HAMPSHIRE RAILWAY.
STATIONS.

^{0}Alexandria^{0}
^{0}Carlinville^{+1}
^{0}Falls Church^{+1}
^{0}Vienna^{+1}
^{0}Hunter's Mill^{+1}
^{0}Thornton^{+1}
^{0}Herndon^{+1}
^{0}Guilford^{+2}
^{0}Farmwell^{+2}
^{0}Belmont^{+2}
^{0}Leesburg^{+2}
^{0}Clark's Gap^{+2}
^{0}Hamilton^{+2}

ALLEGHANY VALLEY RAILWAY.
STATIONS.

^{0}Pittsburg^{+12}
^{0}Lawrenceville^{+12}
^{0}Sharpsburg^{+11}
^{0}Brilliant^{+11}
^{0}Ireland's^{+11}
^{0}Armstrong^{+11}
^{0}Verner^{+11}

Page one of Charles Dowd's system of "plus and minus indexes." The number to the left of each station indicates its zone (numbered 0–3); the number to the right, the local difference, in minutes, from the zone standard. Dowd had this "railway time gazetteer," including indexes for over 8,000 stations on every railway in the country, published at his own expense and sent to railroad managers. (Charles F. Dowd, System of National Time and Railway Time Gazetteer [Albany, N.Y., 1870].)

At train stations, two clocks often told the difference between local and railroad time. Dowd claimed that Buffalo's station included three separate clocks, one for each line using the station plus Buffalo city time. He estimated over eighty distinct railroad times confronting the traveling public in 1870.[50] Though confusing, the situation hardly constituted a crisis, especially if the primacy of local time, sanctified by tradition and the evidence of nature, was assumed. Conductors could call out time differences as they arose, and most cities and towns simply adopted the most convenient local source—usually the time of a jewelry store or factory, as at Lowell; often the railroad time if not too different from the local sun.

The conflicts over the nature of time evident twenty years earlier relaxed in the early 1870s, probably thanks to the kind of local and regional standardization already described at the Lowell mills and at Harvard, Allegheny, and elsewhere. Outside of commercial life, away from telegraphs and train stations, local time ruled the day. Taking national standard time for granted, we can hardly appreciate today how new the idea appeared as late as 1880. Abandoning local times in daily life seemed unthinkable to many—as the Superintendent of the Naval Observatory rhapsodized: "the sun regulates all life upon the earth . . . by local time [man] must live, move, and have his being."[51] Though cumbersome and awkward-seeming to us, if the haphazard system bothered anyone other than a few railroad men and scientists they left little record of their dismay.[52] The potential advantages of standard time, on the other hand, readily impressed anyone stopping to think about it, and between 1870 and 1880 the number of standards used on the railroads decreased by about thirty, thanks to consolidation and cooperation between roads.[53]

Standard time was more than just convenient; it offered a new model for regulating work and public life. The Pennsylvania Railroad's business with Langley certainly did the Observatory a favor—a thousand dollars meant little to the railroad but a great deal to Langley. But what probably intrigued Cassatt most was Allegheny's ready facilities for easy and largely *automatic* telegraphic transmission of a standard time. Langley's system used a small lever just touching one of the toothed wheels of a clock. When struck lightly by the teeth of the gear wheel as the clock ticked, the lever pivoted, regularly making and breaking contact with a telegraph receiving terminal. A subscriber to the time service, Langley wrote, hears what "sounds like a clock ticking somewhere in the room . . . it *is* a clock ticking," he continued, "but the clock is many hundreds of miles away."[54] Allegheny time extended the regulating authority of a single central clock—and by extension, of A. J. Cassatt—over hundreds of miles.

Shortly after securing Cassatt's business, Langley and Thaw began lobbying J. D. Layng, manager of the western divisions of the Pennsylvania, asking an additional thousand dollars a year for time service on the western roads. The Observatory, Thaw enthused, was "a perfect machine for determining that standard time, and communicating it to all needful points."[55] Later that year the two enthusiasts talked Pittsburgh into dropping another thousand dollars per annum to receive Allegheny time on its new City Hall clock.[56] To justify the substantial expense and draw more customers, Langley explained the benefits of his service in a number of letters, addresses, and newspaper and magazine articles.[57]

The most extensive example, in *Harper's Monthly*, made a fetish of accurate observation to sell Observatory time to the general public. Langley described his transit instrument, mounted on massive stone piers "carried down to bedrock" to preserve it from even the slightest vibration. The position of the transit marked the meridian, and tiny cross hairs, made from the "cocoon of the little wood-spider (common cobweb is too coarse)" mapped the regular passage of certain "clock stars" across the instrument's field of view. These stars gave the best indication of the earth's rotation and thus of time. The problem lay in measuring them accurately. Langley described the numerous and pesky sources of possible error facing the astronomer—the stone piers shrank and expanded in heat or cold, shifting the transit to the right or left; the tap of a finger set the transit vibrating hysterically. The axles of the transit, "though turned with a diamond," remained uneven, for "mathematical perfection is not attainable."

Finally, the human observer inevitably introduced a whole new set of errors. Individual reaction time varied from day to day and person to person. Even the delay between the time a star was perceived to have passed the cross hairs and the instant the observer's hand struck a telegraph key, regulating the time service sending clock, counted significantly. "Yet this time must be measured," Langley wrote, "for though we cannot get rid of the faults of the mechanism employed, whether it be of stone or brass, or that of nerves and brain, we can [correct] for their effects, and so finally evolve truth from what appears to be a maze of error."

Langley used machine metaphors to make the observer and the Observatory part of one mechanism. He made "nerves and brain" analogous to "brass and stone." By calling attention to sources of error Langley only increased the reader's admiration for the Observatory's precision and the system of knowledge that kept error in check. Like Adams's defense of astronomy, Langley's time service offered scientific accuracy in place of individual judgment or local custom. But Langley went further than Adams. Adams had looked to the stars as the inspiration for rationality, but Langley sought to make the regulating influence of heavens automatic and unconscious—to bind daily life to mechanical, automatic accuracy.

He described "the immense movement of freight and passenger traffic" roving across the continent. "During the day the rush of passing cars is incessant," he continued, "and when night comes on, the noise of panting engines, which pass the window with a crescent shriek, makes it seem as if our broken sleep had been passed in the vicinity of interminable rushing trains." Hawthorne's "Celestial Railroad" journey, thirty years earlier, took place in a nightmare. Langley's shrieking trains similarly disturbed the nation's sleep, the more so given the vast scale of continental railroad enterprise. Trains entered into everyone's daily life whether they knew it or not, Langley implied, and their effects might not always be wholesome. Scientific *observation* offered a solution to the problems they caused.

"The amount and intricacy of this movement grow on us the more we study it," he mused, and "an eye which could survey the whole at once" would see "two endless processions of trains separated from each other by but a few minutes distance." Langley's imagery, repeated in a number of his other articles, made railroad operations into a kind of national unconscious, a constant circulation taking place even in dreams, vaguely disturbing and fraught with potential danger. The movement of trains defied visual comprehension or the most vigilant human surveillance. Only one thing preserved the system from chaos—"*Time* exactly obtained and kept is the regulator of this complex system of moving parts, which, in theory at least, should resemble one great piece of clockwork." "To make things 'move like clockwork' is not merely a figure of speech," Langley concluded, "here where our lives depend on the accuracy of a conductor's watch."[58]

In his 1856 speech Edward Everett had likened the telescope "to a wondrous Cyclopean eye, endowed with superhuman power, by which the astronomer extends his view to the further heavens."[59] Langley's clockwork metaphor turned the telescope on society. The astronomer extended his view into daily life, using the "superhuman power" of the standard clock and telegraphic signal to regulate social and commercial time. Langley replaced human frailty and propensity for error with automatic, consistent, mechanical observation. His clockwork metaphor replaced Adams's rational electorate with dependent wheels turning in a complex social mechanism.[60]

"The perfect disciplinary apparatus," Michel Foucault remarked in his study of social discipline and criminality, "would make it possible for a single gaze to see everything constantly."[61] Everett and Langley described such an apparatus, an apparatus of standardized time that would oversee and regulate social movements of all kinds. The standard observatory clock transmitted the time, and lesser clocks relayed it to cities, towns, factories, and homes. Standard time offered a framework for social reorganization along new principles, principles of fixed, regular, and automatic control. "Accuracy" bound the framework together and assured its virtue, just as in medieval theology regular schedules of prayer had assured the prayer's piety. Langley posed a new system of social organization and regulating authority propped up by automatic accuracy. The more accurate and precise time signals seemed, the more legitimate their authority over daily life became. Automatically, unconsciously, standard public time offered a new pattern for ordering everyday life and labor.

By 1878, the date of the *Harper's* article, other observatories had begun promoting time services of their own. Harvard's Observatory, for example, started selling time signals in 1871, following Langley's example. A year later the director reported the service losing money, and "its discontinuance seriously considered." But accurate time, the Observatory claimed, was now essential to railroads and their patrons, factories, jewelers, watchmakers, and cities and towns generally. Instead of canceling the service, the Observatory began "active measures . . . to secure a wider knowledge of our time signals, by circulars, letters, and personal interviews."[62]

One such circular issued in 1877, "for distribution among those interested in a common standard of public time throughout New England," stressed again and in considerable detail the accuracy of Observatory time signals and their ease of transmission across vast distances. The circular introduced the Observatory's growing interest and work on behalf of national standard time. Though making no great claims for the advantages of such a time, the circular must have made an impression, since in 1878 the Observatory cooperated with the Signal Service Bureau to place a time ball on the Equitable Life Assurance Company's headquarters. The time ball brought uniformity to Boston's public time.[63]

Now largely forgotten, time balls once played a significant part in the daily life of American cities. We live surrounded by accurate, synchronized time. Computers, office clocks, telephones, radios, televisions, all announce the time. Even the cheapest quartz watch keeps a more regular rate than the earth's rotation, and so we have no special need for assuring the accuracy of any particular timekeepers—there are more than enough to go around, and the framework for standard time is well established. But in the nineteenth century accurate time was a far more precarious quantity. Watches tended to vary, and even the best public clocks drifted away from each other. Time balls provided a free public display of time that checked the accuracy of city clocks and watches. Typically placed on a pole atop some tall tower or readily visible building, the large metal globes usually rose and fell once daily at noon, signaling the time. Ships at anchor regulated their chronometers from a ball visible in port. Watched from an office window, the noon ball called the hungry to lunch. Pedestrians throughout American cities paused a few minutes before twelve, faces turned to the tower, to check their watches. (The time ball survives today to inaugurate New Year's Day from Times Square.)

Harvard's time ball theoretically only standardized Boston time. But the Observatory's actions reflected both increasing interest in national time standards and newly emergent, competing commercial efforts to define public time. As the idea of selling standard time grew more common, competition among time-sellers increased.

In 1877, the Western Union Company had installed a time ball on the tower of its New York headquarters. The New York ball dropped once daily at *local* noon, on a signal received from the Naval Observatory in Washington. The system preserved the time of the New York City meridian, while establishing Western Union's time ball as the most accurate standard of New York time. The telegraph company's particularly cozy relationship with the Naval Observatory had started in 1865. This relationship, redolent of scientific accuracy and objectivity, strengthened perceptions of the telegraph company as a public-spirited, benevolent, and useful institution. Western Union gained the right to distribute free time signals to its offices in return for allowing the Observatory limited access to telegraph wires when official business demanded it.[64] By 1877 the telegraph giant was selling those same time signals— obtained free from the government—to private citizens. Circulars sent to the residents of cities of over 20,000 people began what later became a virtual monopoly on commercial signaling of time. The service placed "controlled clocks" in the "house, office or manufactory" of subscribers, charging between $75 and $375 per year for daily signals. Western Union's time ball bore public witness to the company's concern for accuracy, and had the happy effect of establishing Western Union as *the* standard of time in New York City. By heightening public awareness of time in gen-

The Western Union time "ball," placed atop the company's building at Broadway and Dey streets in 1877, and the mechanism used to hoist the ball to the top of the mast and then drop it automatically at noon on a signal from the Naval Observatory in Washington. (From *Scientific American*, 39, n.s. [Nov. 30, 1878], p. 335. National Museum of American History, Smithsonian Institution.)

eral, the company created an interest in its service and advertised itself as the solution in one stroke.[65]

The Naval Observatory helped promote the system as well. Naval astronomer E. S. Holden, in *Popular Science Monthly*, explained the new service and time ball in detail and praised Western Union's "public spirited" measure "for the benefit of the citizens and the shipping of New York City."[66] Samuel Langley, on the other hand, found the measure an outrage and a threat to the lucrative systems he and other astronomers had been building. The Naval Observatory, he complained to an ally at Harvard, "had not waited for a popular demand but had sent out canvassers so to speak to get business." True, Langley also advertised heavily. But he found the Naval Observatory's public spirit "more mercantile than scientific."[67]

Both time signal systems reiterate the growing public awareness of time and the emergence of time as a commodity, a good to be sold outright or "laid on like gas or water," as Langley put it.[68] Once time seemed nearly indistinguishable from nature and God. Now, thanks to controlling clocks and telegraph lines, the heavens themselves were laid open and mined for profit—time became secularized, a thing identifiable, quantifiable, packageable, and saleable. Formalized in electrical signals, observatory time promised to bring the rationality of science to whatever task lay at hand. Patrons bought the authority and prestige of scientific time through telegraph wires, and applied it where needed to regulate daily life.[69]

Standardized time grew from a regional innovation to a national reform in the 1870s, and "time" became not a matter of local imperatives and individual interpretation, but the subject of bureaucratic wrangling, commercial competition, and scientific backstabbing. The protracted and complex dispute between the Signal Service and the Naval Observatory offers perhaps the best example of the struggle to define public time.

The Army's Signal Service Bureau had long used the telegraph to link regional observatories. Cleveland Abbe, director of the Cincinnati Observatory and father of the Signal Bureau's national weather service, began taking advantage of this network to compile weather bulletins in 1869. Abbe received reports from other observatories and telegraph-equipped weather stations giving conditions at each place. Coordinating the reports through the Signal Service, he translated them into maps with the now familiar "isobars" depicting warm or cold fronts.[70] To Abbe's annoyance, each station, except those kept by astronomers, gave the time of its report by the local standard. Phrases like "railroad time," "local time," or "city time," in the absence of a single standard, offered only ambiguity and confusion. Having pursued standard public time for Cincinnati,[71] Abbe's work with the weather service lead him naturally to the idea of a national standard.

As chair of the American Metrological Society's Committee on Standard Time, in 1875 Abbe began working with other scientists, including Langley and Edward Pickering of Harvard, on a zone system of time based on Greenwich meridians. The Metrological Society, founded in 1873, was a federation of scientists, educators, and other professionals, including F. A. P. Barnard of Columbia, interested in the reform of time and other units of measure and value. It played a crucial role in the adoption of national standard time.[72]

The astronomers also began comparing the meridian times they telegraphed to the Signal Service in Washington. Inevitably, small errors plagued each observatory on occasion. To minimize this variation, they planned to establish a mean standard from the average error at each observatory, unified with the Greenwich time used by much of the international scientific community. Under this "clearinghouse plan," coordinated by the Signal Service, each observatory would be informed of its error from the mean. Local times could be calculated from the clearinghouse standard.[73]

For someone engaged in diverse national meteorological observations, like Abbe, such a time was far more useful if the general public adopted it—otherwise local time standards still clouded the temporal skies. Useful to scientists because calculated from Greenwich, the clearinghouse system's greater accuracy and *international* applicability, according to advocates, made it ideally suited to regulating public time as well. Hoping to make regional time standards a national issue, Abbe's committee began promoting the idea to railroad officials and the federal government in the late 1870s.[74]

The Signal Service, as a first step, had cooperated with Harvard to establish the Boston time ball. It now planned, in 1881, to place a second ball, on the Equitable Insurance Building in New York. But this second time ball would be dropped by the nearest Greenwich meridian, setting up two times for New York City and calling the authority of the Western Union ball into question. The Signal Service hoped, with the help of a $25,000 annual appropriation, to do the same in other cities and eventually standardize American time according to the more accurate and scientifically useful Greenwich signals. Those interested in obtaining the signals were directed to their friendly local observatory and its telegraph time system.[75]

Western Union offered mean local time to its New York customers. The company's time ball kept *New York* noon, calculated from Washington Naval Observatory time. Their circular included a table listing the difference between noon by the Naval Observatory signal and local noon at the meridian of prominent points in cities the service was aimed at. The chart shows the informality of American timekeeping practices—no definitive authority rings the time in any of the cities listed. Instead, there are only prominent public clocks or civic buildings which serve as symbols of the city and its "local time." The service left the question of local time up to the customer; he or she might keep the "controlled clock" on local time or adopt the Washington standard. The Signal Bureau's

TABLE showing the mean time at various places in the United States when it is noon (12h. 0m. 0s.) at Washington, and the annual charges made by the Western Union Telegraph Company for transmitting the Washington Noon Signal to the main office of the Company in such places.

City.	State.	Point to which the Longitude is referred.	Mean time when it is noon at Washington. H. M. S.			Authority	For a Daily Signal.	For a Weekly Signal.
Atlanta	Ga.	City Hall or Court House	11	30	38.61	C. S.	$225	$37 50
Albany	N. Y.	Dudley Observatory, Dome	12	13	12.34	C. S.	150	25 00
Allegheny City	Pa.	Allegheny Observatory, Dome	11	48	9.16	C. S.	150	25 00
Bridgeport	Ct.	Spire, Middle Bridgeport (1834)	12	15	26.42	C. S.	150	25 00
Bangor	Me.	C. S. Astro'l Station on Thomas Hill, near corner Union and James streets	12	33	4.17	C. S.	225	37 50
Baltimore	Md.	Washington Monument	12	1	44.13	C. S.	75	12 50
Boston	Mass.	State House	12	23	56.74	C. S.	150	25 00
Brooklyn	N. Y.	City Hall	12	12	14.28	C. S.	75	12 50
Buffalo	"	Van Duzee's Observatory	11	52	43.81	*	150	25 00
Chicago	Ill.	City Hall or Court House	11	17	40.20	*	300	50 00
Covington	Ky.						225	37 50
Cambridge	Mass	Harvard College Observatory, centre of Dome	12	23	41.11	C. S.	150	25 00
Charleston	"	Bunker Hill Monument	12	23	57.44	C. S.	150	25 00
Chelsea	"	Neptune House Flagstaff	12	24	17.03	C. S.	150	25 00
Camden	N. J.	Church Spire	12	7	43.69	C. S.	75	12 50
Cincinnati	O.	Mitchel's Old Observatory (1848)	11	30	13.05	C. S.	150	25 00
Cleveland	"	Marine Hospital, Dome	11	41	26.39	C. S.	150	25 00
Columbus	"	Dome of Capitol	11	36	13.13	C. S.	150	25 00
Charleston	S. C.	St. Michael's Church	11	48	28.84	C. S.	150	25 00
Davenport	Iowa	Court House	11	5	53.5	A. A.	300	50 00
Dubuque	"	Centre of the City (1859)	11	5	32 4	A. A.	300	50 00
Detroit	Mich	U. S. Lake Survey, New Observatory	11	35	59.88	*	225	37 50
Dayton	O.		11	31	23.—	A. A.	225	37 50

Excerpt from a table describing Western Union's time service. Under "authority," "C.S." means "Coast Survey," "A.A." means "American Almanac." Note the exaggerated accuracy of the mean times, and the remarkably high cost of the signals. (From Proceedings of the American Metrological Society, II [Dec. 1878–Dec. 1879], p. 35.)

New York time ball, on the other hand, would have put Gotham on the nearest Greenwich meridian, approximately four minutes behind New York local time.

Word of the Signal Bureau plan filtered into the press. Abbe himself accused the Naval Observatory of sloppy timekeeping, an *astronomical* sin, in the New York papers, and his comments cast the reliability of the New York time ball into doubt. Angry denunciations of "meddling" by the Naval Observatory followed.[76] To head off the Signal Service program, the Observatory in 1882 sponsored a bill placing time balls, controlled by Washington, on customshouses in the major port cities at taxpayer expense.[77] The bill, Langley sadly assumed, "means the end of our local time services" and thus the Signal Service clearinghouse plan.[78] The bill failed, to Langley's relief. But the clearinghouse idea, never fully established, also collapsed in 1884, when "some evil-minded person" convinced the Secretary of the Navy to suspend Signal Service Bureau timekeeping operations.[79] The "evil-minded person" probably had something to do with Western Union. A falling out in 1874 had put the Signal Service and the telegraph company on bad terms—the Bureau maintaining its right to round the clock transmission of official dispatches, Western Union resisting the Bureau's claims as an attempt to nationalize the telegraph industry.[80] More importantly, if the Signal Bureau succeeded in establishing standard time based on its own measurements, Western Union lost its increasingly lucrative near monopoly on government-derived time signals.

Western Union's house organ, the *Journal of the Telegraph,* ran editorials in 1882 denouncing the idea of standard time for the general public. Although several years earlier the magazine had waxed enthusiastic about the company's New York time ball, now the *Journal* proclaimed standard time contrary to the laws of nature and a tyrannical imposition by "so-called men of science." It ignored local times, they charged, and instead imposed "English time" (Greenwich) on America.[81] With government unwilling to act, no resolution seemed likely. But the terms of the entire disagreement were altered a year later when the railroads intervened, as Chapter Three will show.

The inter-agency squabbling made little impression on the public before 1880. Most probably realized, as the city fathers of Pittsburgh had when considering Langley's hefty bills for time service, that "the consideration of *unity* comes even before that of *accuracy,* and it is not hard to see that if every clock and watch in the community were agreed in being wrong by exactly the same amount, little public inconvenience would result."[82] But the controversy allowed the scientists and businessmen involved to market the idea of standard time to the public.

Advocates of observatory time, again, stressed accuracy, mechanical precision and universal applicability in place of tradition, religious piety, and nature's local example. They found attention to time "strikingly indicative of the character of a people in the scale of civilization." "The increasing requirements of a complex civilization," one astronomer concluded, "demand more and more attention to the keeping of accurate standard time."[83] Astronomers linked standard observatory time to the linear progress of American industry, pointing out that only synchronized, accurate time could regulate business.

Time balls and observatory time signals, advanced as a public-spirited commercial boon, helped make their case. Commenting on Western Union's New York time ball, the *New York Tribune* freely admitted the advantages of standardized time, and praised the improvement in punctuality such innovations encouraged. But the paper cautioned that using Washington time for the entire nation, as Western Union proposed, "would require almost a revolution of ordinary ideas." To preserve local times the editors suggested equipping clocks with two sets of hands, one showing Washington time, the other local time.[84]

The Signal Service astronomers had heard suggestions like that before, and wanted no part of them or of local time. Leonard Waldo, director of the time service at Harvard and later at Yale, claimed in the *North American Review* that "the people have been dictated to by their jewelers regarding their standard of time." "A little reflection shows," he continued, that only a very questionable advantage arose from maintaining a local time "simply because the jewelers of the city insist on a time which shall appeal to the local time of their customers."[85]

Waldo proposed instead having the people dictated to by scientists. "The furnishing of correct time," he intoned, "is educational in its nature, for it inculcates in the masses a certain precision in doing the daily work of life which conduces, perhaps, to a sounder morality."[86] In a report to the railroad commissioners of Connecticut, Waldo referred to the "many thousands of persons" engaged in factories. "Any service which will train these persons into habits of accuracy and punctuality," he claimed, "which will affect all employers and all employees with the same strict impartiality, so far as wages for time employed is concerned, will be a great benefit to the State." Waldo proposed standard time as arbiter and schoolmaster, as the regulating authority for disorganized workers.[87]

By 1880, Waldo had become the most prominent spokesman for national standard time and the Signal Service clearinghouse plan. But his pieties about improving the masses and benefiting the state barely concealed his own profit making. Shortly after leaving Harvard for Yale, Waldo founded a Horological Bureau at Yale's Winchester Observatory. The Bureau was to "encourage the higher development of the horological industries" and promote the idea of standard time by testing and rating watches. Circulars sent to railroads, watchmakers, and jewelers lauded the Bureau's special temperature-controlled vaults and scientific equipment, encouraging corporate clients to send their watches to Yale, with a fee, for examination.[88]

As director of the Horological Bureau, Waldo began lobbying both the local railroads and the railroad commissioners of the state of Connecticut. In March 1881 the state legislature, following Waldo's persuasive recommendations to the railroad commissioners, adopted the time of New York City, signaled from Yale's Observatory, as the legal standard of public time in Connecticut. Their action marked the first example of legislation on public time at the state level, and in intent legally abolished local time.[89]

A year later Waldo formed the Standard Time Company, a joint stock enterprise organized, like Western Union's time service, to provide accurate telegraphed time signals to home and office.

The Standard Time Company enjoyed an exclusive contract with Yale Observatory, sharing its staff and facilities, and offered to spread the bounty of science to all those willing to pay.[90] The phrase "time is money," for Waldo, was now thrice true. He made money as director of the Observatory time service, as manager of the Horological Bureau, and as secretary of the Standard Time Company. The contract with Connecticut made standard observatory time law, increased public attention to accurate time, and brought Waldo business.

Even more than Langley, Waldo made a commodity of time, harvesting, packaging, and selling it to the state and its citizens. Like most advocates of standard time, Waldo based his appeal largely on the ideal of commercial circulation—observatory time facilitated the free movement of goods, services, and information necessary to a modern economy. Connecticut, Waldo reminded businessmen, was essentially a manufacturing center, with "many thousands of persons engaged in her factories upon whose precision and economy of time the success of the State in her struggle for trade must ultimately depend." Observatory time signals, thanks to the prestige attached to their great accuracy, reformed the timekeeping habits of these "persons."

Harvard and Yale's time services, Waldo boasted to the railroad commissioners, "fostered a certain emulation among the employees as to the accuracy and timekeeping qualities of their watches." They acquire a "personal pride in being authorities on the subject of correct time."[91] That is, they internalized the new time and made it part of themselves. Even better, the Connecticut legislature's adoption of New York City time synchronized the state's economy with both the commercial giant to the south, and "that broad expanse of territory between Buffalo and Rhode Island."[92] Within that broad expanse, people and goods supposedly "moved like clockwork," as Langley had put it, synchronized to and improved by the beats of Yale's Observatory clock.

Forty years of wrangling over the nature of time in an industrial society had moved American understanding away from nature and toward the machine. Time had been a quantity invented by God and expressed in nature. Factory schedules, like farmers' almanacs, merely expressed a particular division of an abstract, universal quantity—they described how people were to act within the daily passage of time, according to the demands of the task at hand. They affected not the *sense* of time as much as its *use*.

Observatory time, on the other hand, was something new; a utility, like gas or water, a tool for improving the "morality" of laborers, facilitating the movements of trade, and synchronizing the economies of different regions. It reconstituted the authorities that governed daily life. Technology's "annihilation of space," in the era's ubiquitous phrase, put men and women in distant places in new relations with each other, demanding new understandings of time sensible to commercial operations.[93] Waldo believed, like Langley, that "the railroads ought to be the most useful means in securing uniformity." As the most visible and powerful symbol of both the circulation of goods and money and the necessity for accuracy, railroads were "the great educators and monitors of the people in teaching and keeping exact time."[94]

The railroad's power ultimately did Waldo no good. After 1883, when the railroads adopted a uniform time of their own, Connecticut canceled its contract with the Observatory.[95] The Horological Bureau and the Standard Time Company both folded after the railroads endorsed Naval Observatory time and Western Union's time service.[96] Waldo himself resigned from Yale under a cloud after an inquiry into his fiscal relations with the Observatory.[97] But the railroads largely succeeded where scientists like Waldo failed, establishing the zone system we use today across most of the continent. Railroad time reorganized public time to suit the needs of commerce. In the process it exposed both a deep-seated ambivalence about this new public time and a wide range of objections to its implications.

The Day of Two Noons

American astronomers, drawing on technological innovations like the telegraph, had created a new understanding of time by 1880. The observatories made time a product. They captured an apparently natural phenomenon, distilled what they understood to be its essence — order — and then offered this essence for sale in tidy, attractive packages. Observatories profited financially by selling time, but even more, each hoped to see standard astronomical observatory time replace the confused and by now obsolete system of local times inherited from the past.

But no scientist or scientific society commanded enough leverage to force the idea of standard time on a largely unimpressed public. Time, after all, remained a fairly hazy idea for most people, fraught with vaguely unsettling religious implications and perhaps better left alone. In America in 1880, relatively few people traveled often enough to suffer from the lack of a standard time. Certainly large corporations, especially those doing business by telegraph, felt the lack of a standard time keenly. But for most people, especially outside of cities, "time" and "place" seemed nearly synonymous. In "Up the Cooly," one of Hamlin Garland's short stories on rural life in the Midwest, the protagonist's farmer brother describes how an oppressive mortgage "ate us up in just four years by the almanac." The years passed in poor harvests and almanac pages, not in calendars or clocks. The protagonist's mother, asking about her actor son's life in New York, simply cannot understand "how one could live in New Jersey and do business daily in New York."[1] Away from the world of telegraphs and railroad trains, there was just no reason to care. Only the railroads, the ultimate symbol of commercial expansion, progress, and the conquest of space, had the motive and the power to reform public timekeeping.

In 1883 an association of railroad managers adopted a system of standard time zones basically the same as that we use today. Without benefit of federal law or public demand they managed to rearrange the nation's system of public timekeeping at a stroke, an administrative and public relations coup of impressive proportions. Following the progress of the railroad's innovation reveals some of the implications of the new time. The times before and after November 18, 1883, "the day of two noons," point not to uniformity, but to lingering divisions over time and its meaning. But the measure's practicality in day-to-day railroad operations lent railroad officials an unquestioning belief in its value for ordinary citizens, while the vast economic and political power of the railroad industry assured them of the measure's success.

American railroads traditionally confronted time mostly as a problem of scheduling. Summer's burden of extra passenger traffic, for example, demanded more trains and revised timetables each vacation season. As the railroad network grew, adjoining roads saw a need to coordinate passenger connections and through traffic. By 1872, a large group of railway superintendents had formed a permanent Time-Table Convention, subsequently known as the General and Southern Railway Time Conventions, the American Railway Association, and finally the modern Association of American Railroads. Besides issues of scheduling and running times, the Convention addressed uniform signal systems, coupling devices, and protocols in handling through freight.[2]

Thanks to such cooperation the railroads had evolved a haphazard but practical system of regional times. Most kept a single time across an entire line, usually the time of the largest city they served. Time changes occurred either at breaks between divisions of a large company, or where two different company's lines met. For example, in 1874 the Pennsylvania Railroad bought its time signals from Samuel Langley's Allegheny Observatory. In the East the line used Philadelphia time, but west of Pittsburgh, the breaking point between the eastern and western divisions of the company, trains ran on Columbus, Ohio, time. Smaller lines usually followed the larger, and one line's rescheduling meant headaches for another.

According to its first secretary, William F. Allen, the General Time Convention's semi-annual meetings were intended to "promote harmony by permitting an interchange of views between the superintendents and managers of the various interests." As the only professional organization of railway managers, the group would allow a "more intimate acquaintanceship" between the superintendents of feuding roads, and help smooth over disagreements. But fierce competition between lines, augmented by the Panic of '73 and subsequent railroad rate wars, soured the harmony and kept the meetings rather sporadic. As late as 1882 Allen had to write its president asking if the Convention still existed.[3]

Allen, born in 1846, was a lifelong employee of the railroads who began his career as a surveyor and engineer on the Camden and Amboy line. Rising to the post of resident engineer, Allen helped lay out the company's new track, gaining in the process a thorough familiarity with the day-to-day business of planning and

William F. Allen, "inventor" of standard time. (National Museum of American History, Smithsonian Institution.)

running a road. Allen envisioned railway management, as he put it later in his life, as "a great and wonderful piece of mechanism, strong yet flexible, capable of rapid and powerful exertion, but exceedingly sensitive." In 1872 he joined the staff of the *Official Guide of the Railways and Steam Navigation Lines in the United States and Canada,* becoming the publication's editor a year later.[4]

As its ponderous title implies, the *Guide* (later retitled and hereafter referred to as the *Traveler's Official Guide*), was largely a collection of transportation timetables and schedules. Before standard time, travelers used guides like Allen's to help puzzle out the profusion of times then in use. Allen's *Traveler's Official Guide,* which also listed railway personnel and their offices, served as the "officially recognized standard of reference in the transaction of business between railroad companies," and under Allen eventually replaced the numerous "unofficial" guides it competed against. Allen's work with the *Traveler's Official Guide* — the official catalogue of railway timetables — led naturally to his appointment as secretary of the newly reorganized General Time Convention/Southern Railway Time Convention in 1875.[5]

The Conventions had been hearing about standard time from a number of sources, described in Chapter Two. Besides Langley's contract with the Pennsylvania and his promotional campaigning, Charles Dowd had persisted in bending the railroad ear with his "plus and minus indexes." Dowd got favorable reports from the Convention after addressing it in 1873, and even saw one of his proposals reprinted in the *Traveler's Official Guide* before Allen joined it. But though most railroad men readily admitted the utility of standard time as an idea, few believed it practical. "Circulars on the subject," Allen claimed, "found their way promptly into the pigeon hole or the wastebasket." Competition remained a more interesting diversion, especially since local times seemed firmly fixed by tradition and custom. As one railroad engineer put it, local time was "clearly beyond the power of the greatest power in the land [the railroads] to alter."[6]

But while the railroads sidestepped the subject of multiple times, American scientists and engineers began tunneling into it. Between 1872 and 1882 a number of scientific and professional societies and concerned individuals wrestled with the problem of uniform public time. Often sharing membership and information, most had generally settled on some system of time zones by 1880, following the lead of Cleveland Abbe and the American Metrological Society.[7] But each proposal left the boundaries of the zones largely unsettled, and resisted tampering with the local sun. While nearly all wanted to do away with local time, no individual or group could imagine changing the traditions of centuries.

The director of the Naval Observatory, for example, strongly opposed any tampering with the local sun's authority. He suggested that while Washington, D.C., time could be adopted nationwide for railroads and shipping, custom would demand putting two sets of hands on watches, to tell local and railroad time apart in ordinary life. The American Society of Civil Engineers, led by the Canadian engineer Sandford Fleming, had published a lengthy proposal for world standard time in 1882. The report favored adopting twenty-four world time zones while suggesting some changes in the nomenclature of the hours.[8] Most of the Society's members liked the idea of standard time but feared the change from local custom.

As Brigadier General Montgomery Meigs put it, "if you derange the habits of a people too much they will have none of it." "We travel greatly," Meigs admitted, "but more millions stay at home than go abroad. The housewife keeps the time for the hours of meals and retiring," the general felt; "we men and boys only follow."[9] Meigs associated local time with the home, that familiar, ostensibly tradition-bound bulwark of social order and virtue. In his view, time was a part of the web of inherited customs women spun to give men a sense of place, an anchorage in the rough seas of commercial life.

Echoing Meigs in part, the United States Senate concluded that "it would appear to be as difficult to alter by edict the ideas and habits of the people in regard to local time as it would be to introduce among them novel systems of weights, measures, volumes and money."[10] Meigs linked time with tradition and the cyclical measures of life at home. The Senate report, on the other hand, associated time with mediums of commercial circulation — units of measure and value which, properly reformed, facilitated commerce and industrial expansion while breaking down traditions of particularity and isolation.

Astronomers Samuel Langley and Leonard Waldo had used similar arguments to attack local time. Both viewed local time as an impediment to progress and a hindrance to commercial circulation. By 1880 they had joined, with Edward Pickering of Harvard, the American Association for the Advancement of Science's Committee on Standard Time. The AAAS Committee, like most scientific groups, rejected the idea of one single time for the United States. Such a time "rudely disturbed" the "historical and deeply seated prejudices" regarding local time. Instead, the Committee recommended four time zones, each based on the meridian of a large commercial city. The eastern zone would use New York time, the central, St. Louis time, and so on. The AAAS feared that times based on Greenwich meridians, which as scientists they preferred, would make no sense to the average citizen. Basing time on the nearest large city would unify commercial regions while following the tendencies business exchange and commercial circulation established on its own.[11]

A fourth group, the rather self-importantly named Association for the Reform and Codification of the Laws of Nations, pointed out that "one of the earliest needs of an intelligent being . . . is a means of measuring the lapse of time." Chaired, like the American Metrological Society, by President Barnard of Columbia, the Association claimed that the need for accurate time grew more urgent as civilization advanced. Mechanical clocks marked a high state of civilization, but "social duties and pleasures imply the concurrence of numbers," and an advanced society demanded "a common standard, to which every individual timepiece shall be made to conform."[12]

But ensuring that conformity seemed unlikely. Cleveland Abbe, dubbed "Mr. Isobars" for his work with the weather service, had enrolled William F. Allen in the American Metrological Society in 1879, hoping the railroads could make common cause with astronomers in reforming public timekeeping habits. "Abolish local times," Abbe wrote to Allen enthusiastically, "is the watchword." Allen published, early in 1881, the American Metrological Society's proposals for time reform, and saw to it that the General Time Convention of October 1881 heard reports from Abbe, Barnard, General William B. Hazen of the Signal Service, and the

AAAS. But the reports were largely ignored, and simply referred to Allen for further study.[13]

Even with Allen's help the railroads seemed unlikely to bestir themselves on behalf of uniformity. Most roads had worked out a practical system of regional standards, and as long as profits continued, further cooperation among lines remained unlikely. No one thought the system perfect, but it worked well enough in practice. The last decade had seen a natural decrease in the number of times used, as lines combined or consolidated their positions— from seventy-odd in 1872 to fifty-three in 1882. It seemed reasonable to assume the process would continue, albeit slowly. Complicating matters further, another rate war broke out in 1881 and suspended the Convention. "You represent," Abbe told Allen, both "the desirability" of standard time and "the hopelessness of the situation."[14]

But the situation was not quite as hopeless as Abbe imagined. Western Union's time service attracted publicity, as did Langley's articles. Constant lobbying by astronomers paid off. Thanks to Leonard Waldo's exertions, for example, Connecticut had adopted a uniform legal standard of time in 1882. The measure included no provisions for enforcement, and several railroads ignored it altogether, but it set a precedent for government action on the subject. Almost simultaneously, the Naval Observatory advanced its previously described bill establishing time balls at customs houses and in port cities. The bill and the ensuing public debate, along with Connecticut's legislation, brought standard time into the daily papers.[15]

Growing worldwide interest in uniform time offered another source of encouragement. Other industrializing nations were experiencing the same time problems America did—confusion about timetables, imprecision in the expiration of contracts, lack of synchronization in international scientific work. Increased telegraphic traffic, both national and international, only made the problem worse. In France, "Paris time" competed with railway clocks, which kept "l'heure de la gare," five minutes behind Paris, and local city clocks showing "l'heure de la ville," whatever that might be. German railroads kept to one of five different regional times, while rail travelers kept track of local time by adjusting their watches according to posts set in the ground next to the tracks. Each post alerted the punctual to a ten-minute change in local time. Count Von Moltke of Prussia pointed out that "we have in Germany five different units of time . . . which, since we have become an empire, it is proper should be done away with." Von Moltke warned the German Parliament that confusion over time could greatly impede coordinated military operations.[16]

As in America, most of the impetus for international standardized time came at first from the scientific community. The First International Geophysical Congress, held in Antwerp in 1871, had passed a resolution urging the adoption of Greenwich meridian as the common zero for determining longitude, and the second IGC four years later renewed the resolution.[17] But scientific well-wishing hardly seemed enough to overcome ancient (and modern) political disputes between nations.

Sandford Fleming, chief civil engineer of the Canadian Pacific Railway, had begun pondering the issue of standard time after missing a train connection on a visit to Ireland. Fleming started corresponding with British and American scientific societies in 1876, advancing his plan to divide the world into twenty-four one-hour time zones. But only two years later he was denied a chance to address the British Association for the Advancement of Science on the grounds that his ideas were "too utopian." Many scientists hesitated to shape public policy. George Airy, England's Astronomer Royal, did concede that the question of standard time should be "extensively ventilated by the memorialists." Still, he doubted that the scientific community could dictate public practice.[18]

Fleming began exchanging ideas with Cleveland Abbe in the

spring of 1880, and together their efforts increased discussion of the subject among scientists, mapmakers, and mathematicians. Most recognized the utility of a standard time system, but many wondered why Greenwich should serve as its basis. The Third International Geophysical Congress, meeting in Venice in 1881, heard several new suggestions. Why not use the Great Pyramid of Egypt to mark the prime meridian, asked Scotland's Astronomer Royal? It occupied a central place in Western antiquity, and its shape certainly helped one visualize the idea of an invisible meridian line passing overhead. An Italian astronomer advanced Jerusalem as a likely choice, one with good symbolic potential, while a Swiss astronomer suggested the Bering Strait.

Not all members of the scientific community favored international cooperation. Simon Newcomb, superintendent of the American *Nautical Almanac,* endorsed a standardized time for America. But he met the idea of worldwide standards with hostility. Americans, he declared, "don't care for other nations; we can't help them, they can't help us." Newcomb saw "no more need for considering Europe in the matter than for considering the inhabitants of Mars." But American progress toward standard time, led by Cleveland Abbe, must have encouraged the international scientific community, for in October 1883 the Seventh International Geophysical Congress managed to firmly and decisively recommend the adoption of Greenwich as prime meridian and standard of international time.[19]

Encouraged by international scientific debate, the American Metrological Society had combined with the American Geographical Society a year earlier, to petition Congress with "the names of several thousand signers." The petitioners asked Congress to authorize President Arthur to call an international conference establishing a common longitude and time. The measure was adopted that August, and preliminary inquiries regarding an International Meridian Conference went out in October 1882. The conference would bring together not scientists but diplomats, and take the first steps toward making standard time international law.[20]

The International Meridian Conference would finally meet in October 1884. In what is now Washington's Old Executive Office Building, forty-one delegates from twenty-five countries assembled to hear arguments in favor of standard time and a prime meridian. Greenwich formed the crux of discussion—why should England's Observatory precede all others? Negotiations broke down briefly when the French delegation refused to accept Greenwich over Paris. After some days, the French offered a compromise: they would accept the distasteful Greenwich when the United States and England adopted the metric system, pride of France. But lobbying by delegates from the United States, Canada, and Great Britain, especially Sandford Fleming, carried the day. The International Meridian Conference formally endorsed Greenwich on October 13, with 22 ayes, 1 nay (San Domingo), France and Brazil abstaining in protest.

The conference could only offer its recommendations, not make law, and most countries acted slowly if at all. Germany took up Von Moltke's advice and adopted Greenwich in 1893. France simply chose Paris time as its national standard, and refused to accept Greenwich time until 1911. Even then the proud French fell back on the transparent ruse of defining their standard as "Paris Mean Time, retarded by nine minutes twenty-one seconds"—Greenwich time in all but name. As late as 1905 Portugal, Holland, Greece, Turkey, Russia, Ireland, and most of South America still shunned Greenwich.

Continued uncertainty about time intensified the diplomatic crisis that began World War I. The volume and pace of telegraphic communication outraced diplomats' ability to keep up, and confusion about the precise hour messages were sent only made things worse. But once fighting began, coordinated standard timetables and synchronized watches kept the whole machinery of violence running as smoothly as possible. The war probably did more than any action of scientist or politician to fix world standard time by law, although Holland rejected it until 1940, and the Ayatollah Khomeini deplored its presence in Islam in the 1980s.[21]

Even though in 1882 the International Meridian Conference was little more than a warm invitation from President Arthur, for the American railroads the movement toward world standard time meant trouble. Common practice, as has been described, fixed railroad time changes at the meeting of individual roads or at the eastern and western division of a single company. The resulting jagged, uneven borders made little sense from a scientific point of view, but for the railroads they had the virtue of practicality. Federally established time zones, on the other hand, might divide at inconvenient places, forcing railroad lines to overturn schedules, running regulations, and established custom generally. At least two railroads in Connecticut, for example, found that state's law obnoxious enough to partly ignore, and simply continued to run on Boston time.[22] Early in 1881 Allen published an open letter to his peers in the *Traveler's Official Guide,* requesting suggestions on standardizing time. He received a number of replies, some of which he published a year later, in April 1882, for the benefit of interested railway men.[23] The replies, combined with the legislations cited above, brought Allen to "certain conclusions" which he stated "in informal conversation" at the fall meeting of the General Time Convention in 1882.

Allen's unspecified "conclusions" awakened "a decided interest" at the Convention, where government regulation of the railroads had always given off a bad odor. The delegates asked him to formulate a standard time proposal of his own, and present it at their next meeting.[24] Allen began studying the problem from his position with the *Traveler's Official Guide.* He recognized the utility of using four American time zones, the system most scientists had favored. A zone system deviated the least from local time and its customary association with the position of the sun. What nagged at Allen was the problem of boundaries between zones.

Most time reformers had drawn the boundaries with scientific precision. Cleveland Abbe's American Metrological Society plan, the one most familiar to Allen, suggested four zones exactly fifteen degrees of longitude—one hour—apart. It ignored both state boundaries and the regional time divisions already established in daily railroad practice. Such a plan made scientific sense, but what was science to a trundling freight car? Allen saw no problems with using Greenwich meridians as the zone standards. Practicality ruled the minds of railroaders, and they felt no sentimental or patriotic attachment to the time of one particular city or country. What they wanted most was a plan that altered existing division breaks, and the accustomed day-to-day operations of the roads, as little as possible.

Drawing largely on the existing divisions between lines, Allen proposed a four-zone scheme fitted entirely to the needs of the railways, and presented it privately to a number of managers of the largest roads.[25] With their support arranged in advance, Allen unveiled his plan at the General and Southern Railway Time Conventions in April 1883. He showed delegates a railroad map of the United States with each of the more than fifty standards of time drawn in a different color. The map vividly depicted what Allen called "Hard Scrabble Time," the seemingly chaotic profusion of local standards.

He then revealed a second map displaying the four zones he proposed, the railroads within each zone all drawn in one color. On the second map zone *standards* corresponded to Greenwich meridians, but zone *boundaries* corresponded to existing breaks— part of the eastern and central zones, for example, met at Pittsburgh, where the Pennsylvania road's two divisions already joined and had switched from Philadelphia to Columbus time since 1874.

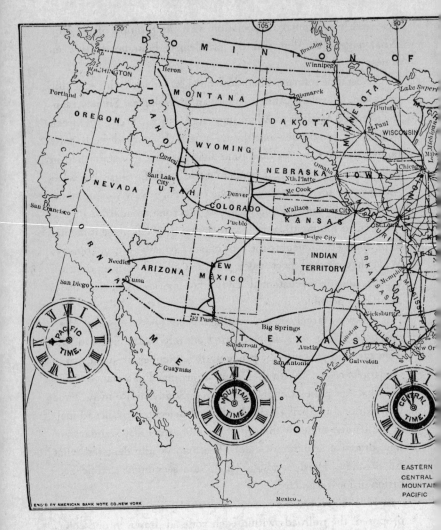

EASTERN
CENTRAL
MOUNTAIN
PACIFIC

Under Allen's plan, few railroads would have to change their schedules or their running times. Most would change only clocks and watches, since the breaks between lines remained largely the same.[26]

"Contrasting the two maps," Allen asked the Convention,

OUTLINE MAP

WITH DIALS SHOWING

Standard Railway Time

PROPOSED BY W. F. ALLEN,

(Secretary Railway Conventions and Editor Travelers'
Official Guide.)

ENDORSED BY GENERAL AND SOUTHERN TIME CONVENTIONS,
APRIL 11 AND 18, 1883,

ADOPTED BY THE CONVENTIONS OF
OCTOBER 11 AND 17, 1883,

AND ORDERED TO TAKE EFFECT ON
NOVEMBER 18, 1883.

TIME   75TH. MERIDIAN
        90TH.   "
        105TH.   "
        120TH.   "

COPYRIGHTED 1883 BY W. F. ALLEN.

William Allen's map of proposed standard time zones in 1883. The map shows how the zones broke at points convenient to the railroad. (National Museum of American History, Smithsonian Institution.)

"which appears more desirable?" The first map, "as variegated as Joseph's coat of many colors," or the second, "with its solid masses of uniform tints?" Joseph's story, of course, told of interfamily jealousy and murder. The old map represented, in Allen's language, "the barbarism of the past" and the pitfalls of divisive

competition. The new map represented "the enlightenment which we hope for in the future." Allen's plan received nearly unanimous approval. It deviated as little as possible from established practice, kept most time changes to half an hour or less, and minimized the adjustments railroads had to make while unifying their running times.

Even more importantly, in presenting his plan Allen insisted that local times, once the railroads had made up their minds, were simply irrelevant—in fact, railroad time would practically abolish

A clearer view of the zone borders as established by the railroads in 1883. (From Scientific American, 17–18 [March 15, 1884], p. 6834. National Museum of American History, Smithsonian Institution.)

them. "What is local time?" Allen asked. "Merely a mean arrived at by calculation and adopted for convenience. For all ordinary business transactions one standard is as good as another so long as all agree to use it." Allen claimed that "railroad trains are the great educators and monitors of the people in teaching and maintaining exact time." The people would take their time from the railroads, no matter what other authorities might deign to intervene, no matter what local traditions prevailed.

"With the standards that cities or governments may adopt,"

Allen continued, "we should have nothing to do." He recalled Connecticut's standard time law and its inconvenience; similar actions, he warned, "may be naturally expected if we do not act for ourselves." Allen voiced all the arrogance of Gilded Age industry when he told the Convention, "we should settle this question ourselves, and not entrust it to the infinite wisdom of the several State legislatures." State legislatures might have their own priorities, scientists their pretty theories. But, Allen insisted, "If we agree that the system here proposed is the one best adapted for practical use on our railway lines, whether it be the best for scientific purposes or not, whether it conforms to the whims of 'the ruling classes' who run our legislatures or not, it is clearly a duty we owe to the great interests intrusted to our charge."[27]

With the Convention's blessing, Allen began sending circulars to the managers of every road, describing the plan and asking whether or not they would accept it. The circulars came back favorably in all but a few cases. Final approval of the plan and the announcement of the date for changing times were scheduled for the Convention's October meeting.[28]

Beneath the brag, Allen and the railroads were worried about state or federal legislation—the "whims," as he put it, of the "ruling classes who run our legislatures." Allen confided to the president of the Michigan Central that he "fear[ed] that nothing short of Congressional action is likely to bring about any uniformity" of time, and that "Congressional action on this subject is to be deprecated." "There is little likelihood of any law being adopted in Washington," Allen continued, "that would be as universally acceptable to the Railway Companies as this movement has proved to be."[29] Government regulation of railroads, with its ominous possibilities for reduced profits, was always a problem. Charged with implementing the change to national standard time, Allen enlisted the most impressive authorities he could find, short of the federal government, on the railroads' behalf.

By basing his zones on the 75th, 90th, 105th, and 120th Greenwich meridians, Allen had already ensured the theoretical endorsement of those scientists pursuing world standard time. Shortly after gaining the Convention's approval Allen wrote to the Connecticut Railroad Commissioners, who had only a year before formulated a legal standard of their own. That state's adoption of New York City time, he declared, was a stumbling block in the way of the new plan — "would it not be possible to have a joint resolution passed . . . endorsing the seventy-fifth meridian line (four minutes slower than New York City time)?" The pliant Commissioners returned their hearty personal approval of the new standard, and the change in law, to the state legislature.[30]

In Boston, however, the 75th meridian differed from local time by sixteen minutes, a fairly substantial change. A number of New England lines agreed to adopt the new time only if the Cambridge Observatory time ball fell by the same standard. Allen replied assuring them that even if the Observatory refused, which was unlikely, the public followed the railroads, not the observatories.[31] But he wanted to make certain. The sort of standard public time Allen campaigned for depended heavily on the imprimatur of scientific authority. Samuel Langley, after all, had sold his time signals on the basis of their irrefutable accuracy, and the Harvard Observatory's endorsement would dispel any taint of intellectual expediency.

Allen wrote to Pickering of Harvard detailing the plan and asking his cooperation. To his alarm, he found Pickering out of the country and an assistant, J. Raynor Edmands, in charge. Edmands expected that Pickering would approve the change, but lacked the authority to approve it himself. Having crawled out on a limb with the railroad managers in private, Allen now needed to avoid a "potentially embarrassing" public failure by selling the eastern standard to the citizens and officials of Boston. If Boston rejected the new time, the whole plan collapsed. Fortunately for

Allen, assistant Edmands burned with enthusiasm for the new time. Loaded down with Allen's maps and the Observatory's prestige, Edmands traveled New England over the next few weeks, addressing the boards of aldermen in Boston and other cities generally. Cleveland Abbe pitched in by writing to President Eliot of Harvard, to assuage Eliot's doubts and present his weighty endorsement to Boston's citizens.[32]

Between Allen's assurances and the scientists' apparent enthusiasms, most New England lines agreed to go ahead and adopt the new standards early, on October 7, 1883, the date of a previously arranged fall schedule change. As it happened, Pickering returned to the country via New York that same day. Edmands met him on the docks, secured his approval, and immediately sent the good news to Allen. Sanctioned by the Observatory, the New England test case proved a limited but encouraging success. Edmands, now joined by Pickering, continued lobbying for public acceptance of the new system. No opposition here, he cabled Allen on a trip to Portland, Maine; "only a lot of cautious men who needed to have doubts satisfied."[33]

But more than a few New Englanders needed their doubts satisfied as word of the railroads' plans spread. "Let us keep our own noon," the prestigious *Boston Evening Transcript* demanded. Referring to Allen's zone boundaries, the editors called the railroads' plan unscientific, since "beginning with a scientific fact, a departure for the convenience of the railroads is made wherever it is deemed wise." Allen's assertion that the railroads controlled public time, moreover, constituted "an impertinence, which is quite of a piece with the familiar claim set up for the railroads that they are the creators of civilization." But in this matter, the railroads *were* the creators of civilization, a new kind of civilization based on standard timekeeping. A letter from "Local Time" urged the *Transcript*'s editors to "expose yet more thoroughly the needlessness and evil of the scheme, and give voice to the general opposition." "Is there no way," "Local Time" plaintively asked, "to prevent the hasty adoption of the scheme by the Fire Commissioners?"[34]

In many cities firehouses supplemented the time ball system, tripping bells or sirens at noon to signal the hour. Boston's firemen had received their time signals from the Cambridge Observatory for many years, and the fire commissioners, after meeting with Pickering and Edmands, had recently come out in favor of the change. As "Local Time" wrote, the Common Council's Committee on Standard Time had added its assent. Rejecting the new railroad standards, the Committee argued, would isolate Boston from the economic life of the rest of the country. Most of Boston's newspapers liked the idea of standard time. The measure awaited only the Council's approval to become law.[35]

But the new standards, the *Boston Daily Advertiser* retorted, "will cause confusion of the worst sort." And "all the confusion and hazard of the change is to be risked, for what?" the paper asked. "Merely to accommodate a handful of people who go between places where there is ten or fifteen minutes difference in local time."[36] The paper had a point—relatively few people traveled that far very often. Little or no popular demand for standard time existed; the change served the railroads above all. Why should Boston have to accommodate the railroads, instead of the reverse? Let them print schedules with local and railroad time, both newspapers declared.

The *Evening Transcript*, and especially the *Advertiser*, spoke for the Boston Brahmin, for the old money that predated Gilded Age industry and industrial time. Standard time struck the editors as yet another modern vulgarity, yet another mark of the relentless industrial stamping mill that crushed out regional distinction and exclusivity. Draped in the raiment of science, the measure in fact represented a cynical disregard for astronomical realities. Boston's time was Boston's particularly, rooted in the history and culture that made Boston different from its southern cousins. Why surrender time to corporate interests? It was in this sense that Reverend William B. White, pastor of the prominent Berkeley Street Congregational Church, rose up and denounced standard time as an immoral fraud, a lie and a "piece of monopolistic work adverse to the workingman's interest."[37]

The fire commissioners, on the other hand, represented new Boston—a rising generation of immigrants, businessmen, and capitalists with few ties to Boston's past and little stake in maintaining its separation. Fire Commissioner Fitzgerald and Alderman O'Brien readily embraced the new time's undeniable practicality and commercial expediency, its tendency to synchronize the economies and markets of eastern cities. The *Daily Advertiser* tried rather feebly to fight fire with fire, claiming that standard time would increase expenditures on gas and electric light.[38] But their opposition cast no new light on the measure's obvious advantages, and seemed out of step with the general tendencies of the city and the age.

"The city, the railroads and the Cambridge Observatory," Edmands cabled Allen, "only await the affirmative action of the Convention before fixing the date for changing all public time in Boston." Read at the fall meeting of the General Time Convention, Edmands's telegram brought a hearty outburst of spontaneous applause from the normally staid managers. On that same day Allen heard word from Leonard Waldo of Yale that his Observatory would also agree to telegraph the new standards to its customers in Connecticut.

Shortly before the Convention Allen had written to the Naval Observatory as well, hoping to gain its cooperation.[39] The Naval Observatory functioned as the quasi-official source of local time in New York, thanks to Western Union's time ball and time service. John Rodgers, Superintendent in 1882, had publicly expressed his objection to abolishing local times on several occasions and in no uncertain terms. Fortunately for Allen, Rodgers was no longer in charge by the fall of 1883. While the new director, Admiral R. W. Schufeldt, pointed out that "we think a single standard would be better, from a scientific point of view," he assured Allen of the Observatory's hearty cooperation. Schufeldt even detailed a junior officer to the Convention to explain how the Observatory's system worked.

This left several ticklish points. The Naval Observatory, a government institution, had agreed to telegraph the new railroad time without legal sanction. No law prevented it from doing this, but then no federal law of any kind had ever been passed regarding standard time. Lieutenant Moore of the Observatory reminded the delegates of the recent bill to provide time balls on customs-houses at Naval Observatory signal. He assured them, incorrectly as it turned out, that the bill was likely to pass in the next session. If it passed, whatever time the Naval Observatory signaled became the national standard.

Whether it passed or not, the railroads planned to present the government with a fait accompli. Once railroads had adopted the new time, they assumed, the railroad's importance to commerce would combine with the measure's universal superiority and practicality to make the new standards indispensable to the public. With the most obvious obstacles to public support now removed, the Convention set the day and hour for the change: twelve o'clock noon on Sunday, November 18, 1883.[40]

As the day drew near Allen began worrying about the Naval Observatory's partner, Western Union. The telegraph company had initiated a time service, described in Chapter Two, based on Naval Observatory signals, and had dropped a time ball daily at New York local noon since 1877. The company sold *local* time. "You are probably aware," Allen wrote the president of the New York Central, "that the Western Union time ball is connected by wire with the workshops of all the principal watchmakers of this city." If Western Union changed its time, Allen admitted, the watchmakers, and thus the watches, of the entire city would follow.[41]

Allen informed General Eckert of Western Union that "your assurance in this matter will materially aid in securing the adoption of the plan." Contemplating a standoff between two of the most powerful industries in the country, Allen cautioned Eckert, "I think that you will find it both practical and desirable to acquiesce in the suggestion."[42] But in private meetings with Eckert and his assistant, General Superintendent Bates, Allen obtained only luke-warm assurance of cooperation from both.

To kindle their enthusiasm, he wrote to Charles Pugh of the Pennsylvania Railroad. "I will arrange to have some little pressure brought to bear upon Messrs. Eckert and Bates," Pugh returned. "Rest assured that all enterprises of this character will promptly adopt the standard after it has been thoroughly established by the railroad companies." Encouraged by Pugh's intervention, telegrapher Bates's regard for standard time soon warmed enough to invite Allen and his family into the company's operating room on November 18, where they could watch the ball drop at the new time.[43]

With help from James Hamblet, supervisor of Western Union's time service, Allen converted Mayor Edson of New York to the new standards on October 19. Allen and Hamblet wrote to several other city departments, but as in Boston the final success of the plan rested on the work of astronomers. J. K. Rees of Columbia, another member of the American Metrological Society, approached the board of aldermen, who resolved to change New York's time with the railroads on November 18. A number of interviews and addresses by Rees and F. A. P. Barnard helped persuade the public, while New York newspapers added their nearly unanimous approval. Allen had managed to remove the most important formal obstacles to the new plan.[44]

Sunday, November 18, the "day of two noons," came and went with hardly a hitch. Each railroad line issued detailed orders on making the change, many of which Allen preserved. The Baltimore and Ohio Railroad, for example, ran its trains on Columbus time, twenty-eight minutes faster than the new central zone standard. At precisely 12:28 p.m., "all Conductors, Engineers and others concerned," orders read, "must be on hand at a telegraph station" to receive the new standard and set their watches back twenty-eight minutes. Those not near a telegraph station simply followed train conductors as they set their watches back. Trainmen on the road set their watches back, then proceeded cautiously to the first station to verify the time.[45]

In most cities people gathered at jewelry stores and near public clocks, waiting expectantly to see what happened. After all, the *New York World* pointed out, "the number of people who understand what this innovation is" was fairly small.[46] Crowds of several hundred began forming in front of New York's Western Union building as early as 11:30 a.m., to await the time ball's drop. In Boston a similar crowd of about two hundred waited with "a sort of scared look on their upturned faces."[47] Several papers claimed that immigrants especially had trouble understanding what went on. The *New York Times* described two stock Irishmen, supposedly watching in puzzlement as the clock stopped, waited, then started again. "Begorra," remarked one in newspaper Irish, "Divil a change at all, at all I can see." The pair walked away in disgust.[48]

A surprising degree of uneasiness attended the transition to the new time, even among those who approved it. Americans could find no precedent for such large-scale tampering with clocks, and the fetish of accuracy that buoyed up the new standards heightened anxiety about the change. In Chicago the stationmaster hovered over his chronometer with a powerful magnifying glass, while anxious railroad employees gathered round. "All looked unusually solemn, and their faces showed that something of an extraordinary nature was about to happen." After the appointed hour came and went, "everybody departed with a sigh of relief." Other newspapers reported a similar sense of relief when the change finally happened.[49]

Standard time posed a new authority for time, a new definition of its source. Newspapers frequently compared the railroad's innovation to Joshua's miracle at Jericho, and though they meant it in a spirit of fun the comparison made sense—in each case the authority for time bent to the will of social forces. *The Washington Post* called the new system of timekeeping "scarcely second to the reformation of the calendar by Julius Caesar, and later by Pope Gregory XIII." The *Post* was right in pointing out that standard time represented the triumph of new priorities for social organization. But it felt obliged, tongue in cheek, to reassure

Circular issued by the Yale College Observatory alerting employees and time service subscribers to the time change. (Courtesy of William F. Allen papers, Manuscript Division, New York Public Library.)

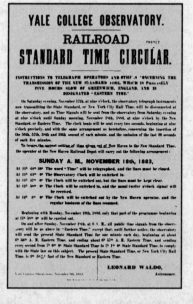

LOUISVILLE & NASHVILLE RAILROAD CO.

OFFICE OF THE GENERAL SUPERINTENDENT OF TRANSPORTATION

Louisville, November 9. 1883.

CIRCULAR No 80.

Important Notice--Change of Standard Time.

TO TAKE EFFECT ON SUNDAY, NOVEMBER 18, AT 10 O'CLOCK A. M.

8897

On Sunday, November the 18th, at 10 o'clock a. m., the **standard time of all Divisions** of this Road will be changed from the present standard, Louisville time, to the **new standard ninetieth meridian or central time,** which will be eighteen minutes slower than the present standard time.

The system to be adopted in changing **Regulators, Clocks, and Watches** will be as follows:

On **Saturday, November** the 17th, at the usual hour for sending time—namely, **10 o'clock a. m** —all **Clocks** and the **Watches** of all **Employes** must be set to the exact present **standard time. On Sunday, November** the 18th, all **Telegraph** Offices must be open, and all Operators on hand for duty not later than 9 o'clock a. m., and remain on duty until relieved by the Chief **Dispatcher** of the Division.

All **Work Engines** and Crews, Road Masters, Supervisors, Section Foremen, and all other **Employes** who are required to have the correct standard time, must report at their nearest Telegraph Station not later than 9:30 a. m., on Sunday, November the 18th, and remain at the Telegraph Office until the **new standard time is received,** and their Watches are set to the correct **new** standard time.

On **Sunday, November** the 18th, the present standard time will be sent over the wires as usual, at 10 o'clock a. m., and as far as possible Dispatchers must have **regular trains at Telegraph** Stations at that hour.

At Precisely 10 o'clock a. m., by the present standard time, all **Trains and Engines,** including **Switch Engines,** must come to a **STAND STILL FOR EIGHTEEN MINUTES,** wherever they may be, until 10:18 a. m. by the present standard time, and at **precisely 10:18 a. m.,** by the present standard time, all **Clocks,** and the **Watches** of all employes must be turned back from 10:18 to exactly 10 o'clock, which will be the new standard time, and the new standard time will then be given from Louisville over all Divisions of the road.

Bulletin alerting employees of the Louisville and Nashville Railroad to the change to standard time on November 18, 1883. This particular line required all trains in motion on that Sunday morning to STAND STILL FOR EIGHTEEN MINUTES. (Courtesy of William F. Allen Papers, Manuscript Division, New York Public Library.)

those "who may imagine that this shaking up of time standards may endanger the stability of the universe" that the sun, moon, and stars would continue to move as they always had.[50]

Most newspaper reporters were unable to pin down just what the change in time meant, and their humorous musings on the subject betrayed their confusion. The *Indianapolis Sentinel,* for example, asked puckishly if on November 18, a man slipping on a banana peel would have sixteen minutes more time to fall before hitting the ground, and if in that time he might secure a mattress to cushion the impact. The *New York Times* included a poem asking "Presidents! Train starters valorous / Astronomers also of all degrees . . . makers of Time Tables rigorous / dispatchers quick as the darting bees" to "Give us the minutes you've snatch'd away!"[51]

Though most cities adopted the new time readily and with little trouble, in Chicago confusion reigned for a few days—city officials as well as a major commuter line and one of the streetcar systems refused to make the change. Allen had done no lobbying in Chicago and failed to secure the Dearborn Observatory's assent. Those holding time service contracts with Dearborn, including the city government and most major jewelers, were reluctant to adopt a time other than that they had paid for. But as other cities and the Observatory adopted the measure, Chicago fell into line. The clock showing "New York Time" came down from the wall at Union Station, and a sign declaring "Philadelphia Time" over a second clock disappeared beneath the legend "Standard Time."[52]

A few wily businessmen incorporated the new time into their advertisements. "Standard time," a Maine druggist reminded customers, "gives us sixteen minutes more in which to do our shopping." A major New York jewelry firm asked Allen for copies of his new map, to sell while promoting themselves as "Keepers of

City Time." An Atlanta jeweler followed a picture of a smoking locomotive with an offer to put a second set of "gold railroad time" hands on any watch. And a St. Louis clothier blared "T-I-M-E! We think the change from local to standard time a most excellent and practical scheme—the future great don't lose any time by it either." Then the tag line: "as the time of year has arrived when warmer attire must be provided. . . ."[53]

Newspaper and magazine editors especially liked the new time. The *Atlanta Constitution* commented on "the utter contempt into which the sun and the moon have fallen. Formerly the rising and setting of the sun was the standard of time in all localities, and in the country the almanac and the sun together enabled many a farmer to do without his watch or clock."[54] Under standard time, the editor pointed out, the sun rose twenty-two minutes ahead of the almanac's prediction. While confessing some misgivings, the paper called the new system of time "the blue line standard on which Atlanta will progress into the future."[55]

The Atlanta paper recalled the pervasive tension between cyclical and linear time. If the almanac and the sun in concert represented cyclical time, then standard time, as the writer pointed out, seemed to emphasize the linear side of western thinking about time—the "blue line standard" of "progress into the future." The phrase repeats the transition from Rittenhouse's orrery to Allen's railroad locomotives. An orrery was a machine that reproduced celestial mechanics—it depicted the movements of the planets, and cyclical time. Like the orrery, clockwork wheels reproduced cyclical time. But clocks also recalled inexorable linear progress into the future. A steam locomotive, with its cyclical, clocklike mechanism of gears, recalled the orrery's complex of parts. But a locomotive *moved*; geographically across space, and forward into the industrial future—a locomotive orrery, a moving clock. Standard time governed this progress. As the accuracy, the precision, of a locomotive's parts regulated and ensured the safety of its progress, standard time regulated, ordered, and ensured the safety of American society's industrial future.

A Nebraska editor, punning unintentionally, called standard time "a striking event" in "our progressive civilization." Its invention counted as "signal proof" that the railroads "are the most potent factors of that civilization." Allen had needed to do little or no campaigning for standard time in the West. In many cases the railroad was probably so important that its example went unquestioned, and in less populous states, settled comparatively recently by whites, there may have been no long-established patterns of life to upset. The Nebraska editor praised the railroad's new time for giving structure to progress. Early American political theorists had looked at the clock and seen a model for rational government. Partisans of standard time looked at the railroad's innovation and saw an intellectual/mechanical device for social regulation. "In a quarter of a century," the Nebraska railroad partisan hymned, "they have made the people of the country homogeneous, breaking through the peculiarities and provincialisms which marked separate and unmingling sections." Standard time, a virtual "revolution," apparently completed the process of homogenization the railroads began.[56]

Accustomed as they were to gathering accurate information from distant places, newspaper editors praised standard time's imposition of uniformity. "Had there been stretched across the continent yesterday a line of clocks," the *New York Times* wrote, "there would have been a continuous ringing from the east to the west." "Half a million clocks," a Pittsburgh daily enthused, "were disturbed at the same instant." "Railroad time everybody's time," declared a headline in St. Louis, while even the cranky *Boston Daily Advertiser* praised the fact that "all laborers who wait until the clock strikes seven a.m. before they strike in with the pick and shovel, will strike together" under standard time.[57]

While mistakenly crediting Charles F. Dowd for the change,[58] *Harper's Weekly* concurred—after standard time, "all the clocks on the continent struck together." Formerly, in respect to time, "the whole country was a pathless wilderness." The poor traveler's watch "was to him but a delusion; clocks in stations star[ed] each other in the face defiant of harmony," striking like "incoherent cowbells in a wild wood." Now, thanks to standard time, "the minute hands of all were in harmony with each other."[59] Like the newspapers, *Harper's* celebrated the apparent fulfillment of Samuel Langley's vision, an industrial society made to "move like clockwork." The widespread and genuine success of standard time gave the impression of a unified, synchronized, and rational society approaching the future in an orderly fashion.

"The system adopted by you," Allen boasted more or less accurately to his fellow railroaders at the next Convention, "now governs the daily and hourly actions of at least fifty million people." Strengthening Allen's case, the International Meridian Conference, meeting in Washington in 1884, voted to recommend a system of Greenwich meridian zones for the world. Despite France's gallant resistance to time based on an English standard, World Standard Time began making slow headway in the international scientific community almost immediately.[60]

Standard time altered relations between individuals and society. In 1883 William Graham Sumner, America's leading social Darwinist, had described the transition from a society based on local traditions of status to a society based on contract. "In the Middle Ages," Sumner wrote, "men were united by custom and prescription into associations, ranks, guilds, and communities of various kinds." These ties were lifelong, and thus "society was dependent, throughout all its details, on status, and the tie, or bond, was sentimental." In modern America, however, "more than anywhere else, the social structure is based on contract." Contract relations are "rational . . . based on sufficient reason, not on custom or prescription." Standard time was fundamental to relations of contract, which required uniform, unvarying units of time. And like contractual relations it was based on nothing more than convenience. Under standard time, men and women governed themselves not by some "sentimental" moral criteria drawn from nature, but by business expediency.[61]

Allen's triumph, and his claim of standard time "governing" daily life, bore out his belief in the power of the railroads—commerce, in effect, now made up its own rules regarding what time was and how it should be understood. Even the venerable *Old Farmer's Almanac* acknowledged the railroad's influence—and changing times—by offering translation tables to help farmers figure the new time against local meridians. In a final insult to solar authority, Philadelphia's park commissioners debated whether or not to adjust the city's sundials into harmony with the new time.[62]

But Allen's innovation, for all his effusive rhetoric, never enjoyed quite the popularity he claimed. Though he omitted much of it in his accounts, from the beginning there was opposition—decidedly in the minority, but vigorous and lasting opposition nevertheless. The problems arose most often, not surprisingly, in places where standard and local time differed more than a few minutes. Residents of these places experienced not just a few minutes' change on the face of a watch, but a dramatic variation in sunlight—in some cases more than half the effect we gain today from daylight saving. Such a difference highlighted the arbitrary nature of the new time, and brought objections on the grounds of legality, hostility to the railroads, regional pride, and religious truth. Boston provided an early example.

Many Bostonians, as described above, were already some-

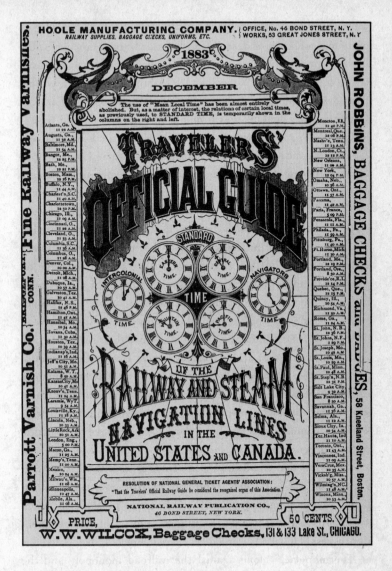

TRAVELERS' OFFICIAL GUIDE

STANDARD

EASTERN TIME. CENTRAL TIME.

INTERCOLONIAL TIME. **TIME** NAVIGATORS TIME.

MOUNTAIN TIME. PACIFIC TIME.

TIME.

OF THE RAILWAY AND STEAM NAVIGATION LINES IN THE UNITED STATES AND CANADA.

Atlanta, Ga. 11 22 A.M.
Augusta, Ga. 11 33 A.M.
Baltimore, Md. 11 54 A.M.
Bangor, Me. 12 25 P.M.
Bath, Me. 12 21 P.M.
Boston, Mass. 12 16 P.M.
Buffalo, N.Y. 11 44 A.M.
Charlest'n, S.C. 11 40 A.M.
Charlottetown, 12 50 P.M.
Chicago, Ill. 11 09 A.M.
Cincinnati, O. 11 22 P.M.
Cleveland, O. 11 33 A.M.
Columbia, S.C. 11 36 A.M.
Columbus, O. 11 28 A.M.
Denver, Col. 10 00 A.M.
Detroit, Mich. 11 28 A.M.
Dubuque, Ia. 10 57 A.M.
Galveston, Tex. 10 41 A.M.
Halifax, N. S. 12 46 P.M.
Hamilton, Ont. 11 42 A.M.
Hannibal, Mo. 10 54 A.M.
Havana, Cuba, 11 30 A.M.
Houston, Tex. 10 39 A.M.
Indianap's, Ind. 11 16 A.M.
Jeff'n City, Mo. 10 51 A.M.
Kalama, W.T. 8 50 A.M.
Kansas City, Mo. 10 44 A.M.
Knoxv'e, Tenn. 11 24 A.M.
Laramie, W.Y. 9 54 A.M.
Louisville, Ky. 11 18 A.M.
Lincoln, Neb. 10 33 A.M.
Little Rock, Ark. 10 51 A.M.
London, Eng. 5 00 P.M.
Macon, Ga. 11 25 A.M.
Memp's, Tenn. 11 00 A.M.
Mexico, 10 24 A.M.
Milwau'e, Wis. 11 08 A.M.
Minneapolis, 10 47 A.M.
Mobile, Ala. 11 08 A.M.

Moncton, N.B. 12 40 P.M.
Montreal, Que. 12 06 P.M.
Nashv'e, Tenn. 11 13 A.M.
N. London, Ct. 12 12 P.M.
New Orleans, 11 00 A.M.
New York, 12 04 P.M.
Omaha, Neb. 10 36 A.M.
Ottawa, Ont. 11 57 A.M.
Panama, 11 42 A.M.
Paris, France, 5 09 P.M.
Pensacola, Fla. 11 13 A.M.
Philad'a, Pa. 11 59 A.M.
Pittsburg, Pa. 11 40 A.M.
Ft. Huron, Mich. 11 30 A.M.
Portland, Me. 12 19 P.M.
Portland, Ore. 8 50 A.M.
Provide'ce, R.I. 12 14 P.M.
Quebec, Que. 12 17 P.M.
Quincy, Ill. 10 54 A.M.
Richmond, Va. 11 50 A.M.
Rome, Ga. 11 20 A.M.
St. John, N. B. 12 36 P.M.
St. Johns, N.F. 1 29 P.M.
St. Joseph, Mo. 10 42 A.M.
St. Louis, Mo. 10 59 A.M.
St. Paul, Minn. 10 48 A.M.
St. Step'n, N.B 12 31 P.M.
Salt Lake City, 9 32 A.M.
San Francisco, 8 50 A.M.
Savannah, Ga. 11 36 A.M.
Selma, Ala. 11 13 A.M.
Sioux City, Ia. 10 34 A.M.
Ter. Haute, Ind. 11 10 A.M.
Toronto, Ont. 11 43 A.M.
Vincennes, Ind. 11 09 A.M.
VeraCruz, Mex. 10 35 A.M.
Vicksb'g, Miss. 10 57 A.M.
Wilming'n, N.C. 11 48 A.M.
Winona, Minn. 10 53 A.M.

Cover of William Allen's Travelers' Official Guide *for December 1883.*
(National Museum of American History,
Smithsonian Institution.)

what hostile to the new standards. Shortly before standard time one Horace Clapp, a debtor, had been ordered to appear before Edward Jenkins, Commissioner of Insolvency, between nine and ten in the morning of November 19. Debtor Clapp appeared that day at what he thought was 9:45 eastern time, only to find that Commissioner Jenkins was running the court on the old time — it was 10:01, and Clapp was in default. Clapp sued Jenkins, and the case wound up before Massachusetts Supreme Court Justice Oliver Wendell Holmes. After several weeks' deliberation Holmes proved true to legal pragmatism. Though no law specifying the legal standard then existed, Holmes ruled that standard time, by its general adoption, was the time in effect when Clapp appeared.[63]

Holmes's decision, backing the authority of an innovation as yet unsanctified by law, strengthened Allen's case, since the legality of standard time troubled many. A number of communities in Maine were sharply divided over the same issue, most prominently the small city of Bangor. When Bangor's City Council voted to adopt the new standards, the mayor vetoed their resolution. He found the change of time involved, almost twenty-five minutes, needless and confusing. "Philadelphia time," as Mayor Cummings called it, seemed senseless in Maine. Any innovation "that disarranges the customs of a people handed down from time immemorial," Cummings concluded, "should at least have the sanction of State legislature."[64]

The mayor's stand drew jeers from around the nation. Advocates of standard time, especially railroad journals, scorned poor "Mayor Dogberry" (Shakespeare's fool in *Much Ado About Nothing*) as an old-fashioned, foolish, and shortsighted impediment to progress. He stands, the *Boston Journal* mocked, like "Leonidas at Thermopylae," alone against the railroad monopoly and the "centralizing tendency of the age."[65] By elaborately spoofing the mayor's rather reasonable objections, and identifying him as typical or representative of objections to the new time, partisans of

Allen's innovation made their opponents seem like cranks, hicks, and relics of the past. Since most major cities had adopted the new time, *Railway Age* asked, "why should any other places, big or little, be allowed to retain the obsolete fashion? Local time must go."[66]

Voters in Bangor felt otherwise. A referendum on standard time in March 1884 disclosed that better than three fourths of the city opposed the railroad innovation, especially among the working class. Opposition grew statewide—despite the public scorn heaped upon Bangor, the Augusta, Maine, *Journal* reported in 1884 that "the cities of this state, one after another, are returning to local time." Anxious to quell any opposition to the new standards, Allen worked privately to swing Maine into line. He wrote letters to the New York papers denying the collapse of standard time in Maine, while privately urging railroad officials and the editor of the *Maine Farmer* to promote the reform more aggressively. With their help Allen shepherded a standard time bill through the state legislature in 1887, despite the continued reluctance of citizens in Bangor and a number of other cities.[67]

Though Allen downplayed it, in fact the range of opposition to standard time was remarkably wide, and confined neither to rural people or small towns. In Washington, Attorney General Benjamin Brewster issued an order forbidding government offices to adopt standard time until Congress authorized it. Like Mayor Cummings, Brewster also became a figure of fun—a common apocryphal story had him rushing to catch a train, only to stand chagrined at the platform as it pulled away by the new time. Congress adopted eastern time for the District of Columbia in March 1884, thus answering Brewster's objections. But the federal legislature largely avoided the issue of national standards.[68]

"Today the 75th meridian is standard, because the railroad kings have so ordered it," declared one congressman, hoping to establish a legal standard. "Tomorrow the railroads may make it the 76th, or the 80th." Why not have a time standard determined by the people? he asked. One of his statesmanlike opponents demanded instead that Congress legalize the new railroad time, since "it is never too late," as he put it, "for the Representatives of the people to catch up with the business interests of the country." Posed as an issue of the railroad's power, the bill died in Committee, and Congress left the matter undecided until 1918. Time seemed like a difficult subject to legislate, especially if the railroads rushed in where politicians feared to tread.[69]

In the absence of federal law, many cities attacked the new time. The *Louisville Courier-Journal* called it "a compulsory lie," "a monstrous fraud" and "a swindle." Can you tell me, a letter asked the editors, "if anyone has the authority and right to change the city time without the consent of the people, and what benefit Louisville can derive from it?" The editors responded that no such authority ruled, and no benefit seemed likely from what was "only a disguised step towards centralization . . . a stab in the dark at our cherished State's rights. After they get all our watches and clocks ticking together," the editors asked in reflexive alarm, "will there not be a further move to merge the zone states into districts or provinces?"

The 90th Greenwich meridian set Louisville back almost eighteen minutes. Why, the paper asked, "should we concede to John Bull's dull Greenwich the position of time dictator? Now what is Greenwich to us? A dingy London suburb." In June 1884 Louisville dropped standard time, and though it would be readopted by the city over the next years, as late as 1905 the Supreme Court of Kentucky could point out that "sun time" was still in use among the bankers of Louisville. According to the Court the word "noon" had no legal meaning in Kentucky except as its usage in a given community described — railroad time could not be considered the legal standard.[70]

Without legal guidelines, cities on the new time zone borders, like Pittsburgh and Wheeling, West Virginia, were just "betwixt and between," as the *Wheeling Daily Register* put it. Wheeling wavered between the eastern or central zone; one moved the clocks ahead twenty-three minutes, the other back thirty-seven. The city waited four years before adopting eastern time. Pittsburgh rejected standard time until 1887. "All the people have to say on it," a Pennsylvania railroad superintendent wrote to Allen, "is that they have always used local time and that it would confuse them to change." "God Almighty," one Pittsburgher exclaimed, "fixed the time for this section just as much as he did for Philadelphia or New York."[71]

In Georgia, bisected by Allen's eastern and central zones, the change proved even more confusing. Augusta, on the border of two zones, opted at first for eastern time, while Savannah and the rest of Georgia were put on central. Assigned by Allen only to accommodate existing railroad division breaks, the partition sat poorly with many Georgians, and both cities soon dropped standard time.[72] Although for a time Augusta moved into the central zone, the problems returned when Augusta and Savannah adopted eastern time in 1888. Even then opposition lingered. "The only people that want standard time are the railroad people," declared a Savannah alderman. "The city doesn't want it, the laboring people don't want it . . . nine tenths of the population of the city are opposed to it." "And the ladies," the courtly alderman added, "why, they have always been opposed to it."[73]

Georgia's hostility to standard time was more cogently phrased by its Supreme Court, ruling in 1889 on whether standard or local time prevailed in the expiration of contracts. "To allow the railroads to fix the standard of time," the Court proclaimed, "would allow them at pleasure to violate or defeat the law." "The standard of time fixed by persons in a certain line of business," the decision continued, "cannot be substituted at will . . . for the statutes of the state, as well as the general law and usage of the country." The Georgia Court clearly saw standard time as a measure limited only to the railroads and having no binding authority.[74]

The railroad's presumption of authority angered and puzzled the Indianapolis *Sentinel*. Labeling standard time "a revolution . . . anarchy, chaos," the *Sentinel* sarcastically added, "we presume the sun, moon and stars will make an attempt to ignore the orders of the railway convention, but they will have to give in at last." People would have to marry by railroad time, eat, work, be born, and die by railroad time. The Indianapolis editors resented railroad time's—and the clock's—displacement of the sun. We compare the city clock with our almanacs, a second Hoosier complained, only to find that "the city clock is a moving falsehood." Which authority prevailed? Could someone really be born by railroad time?[75]

In Charleston, South Carolina, "A Planter" claimed that "country people, not having access to the general press, have failed to see the essential points." "By whose authority," he asked in perplexity, "was the variation [in time] made?" The Natchez, Mississippi, *Daily Democrat* simply if rather confusedly declared, "we fail to see the necessity for the new sidereal time business. The old way of time has answered the purpose for several centuries past, and we think it is about as good a system as can be inaugurated."[76]

No state objected to railroad time as strenuously as Ohio. The *Cincinnati Commercial Gazette* especially objected to adjusting clocks back twenty-two minutes. "The proposition that we should put ourselves out of the way nearly half an hour from the facts so as to harmonize with an imaginary line drawn through Pittsburgh," railed one of a series of editorials, "is simply preposterous . . . let the people of Cincinnati stick to the truth as it is written by the sun, moon and stars." The *Commercial Gazette* thought it "a great stupidity" to assume that people would "accommodate themselves to the railroad convenience." Up to 1890, it published railway timetables under the proud heading "*This is Cincinnati Time. Twenty-two minutes faster than railroad time.*"[77]

While other Cincinnati papers favored the new standards, a number of Ohio cities shared the *Commercial Gazette*'s hostility. A Dayton City councilman failed to see "why Dayton should keep railroad time—they don't even recognize the city, nor confirm their time to suit Dayton's, no more should Dayton change to suit them." The *Columbus Dispatch* reported that after the change, "jewelers were flooded today with the watches of customers left [to have] a third hand put on." One jeweler received more than sixty. Columbians saw little reason to give up local time. After briefly embracing the central zone, the City Council officially abandoned it a week later. Asked if his city would adopt the new time, Toledo's mayor replied, "I do not see why we should. The difference is so great that . . . it would result in no little inconvenience." Akron's *Daily Beacon* warned of confusion "if some of these illogical enthusiasts" of standard time "are not muzzled."[78]

Opposition reached a height of some kind in 1889, when Bellaire, Ohio's, Board of Education attempted to set the schoolhouse clock ahead to eastern time. City Council, objecting to the change, had the entire board arrested. Proud of Bellaire's action and deploring any attempt to legislate a state-wide standard time, the county newspaper declared, "if any members of the Legislature are anxious to set their watches at what they call 'Central Standard Time' they should forthwith do so; but when some crank among them proposes to enact that time into law—to correct and regulate the sun . . . he is carrying the joke a little too far."[79]

But a local court upheld the Board of Education's right to regulate its schoolhouse clock, leaving Bellaire with three times: the "true time" of the sun, "Philadelphia [eastern] time" on the schoolhouse, and "St. Louis" [central] time at the railroad station. Bellaire's rather silly dilemma recapitulated the widespread and diverse popular hostility to standard time in Ohio. Neither an isolated rural phenomenon nor a pet project of cranks, Ohio's hostility denied the railroad's claims of temporal hegemony. Back in 1884 Allen had written to railroad managers asking for lists of those communities rejecting standard time. In Ohio the list included Cleveland, Columbus, Springfield, Dayton, Cincinnati and neighboring Muncie, Indiana, each a substantial city and important manufacturing center.[80]

Standing midway between two meridians, Ohio saw no advantage to adopting any other region's time. A letter to the *Commercial Gazette* suggested that if zone times must be adopted, a separate "Ohio zone" halfway between eastern and central time would better serve the public. Detroit expressed the same reluctance, and despite statewide adoption of central time in 1885, Detroit changed standards repeatedly well into the twentieth century. The difference between railroad and local time, wrote young Henry Ford, then amusing himself by fixing watches in his native Dearborn, Michigan, "bothered me a good deal." Ford designed a watch with two dials, keeping local time separate from the railroad's. The region's hostility to standard time led Charles Dowd to work up a plan calling for a similar "intermediate zone" between central and eastern time.[81]

Partially consoled by the fact that in Ohio "the traveling public, at least, will have to eat, drink and sleep by standard time," Allen welcomed an offer of assistance from Professor T. C. Mendenhall of the Coast and Geodetic Survey. Arming himself with copies of Allen's maps, Mendenhall addressed the Ohio Society of Civil Engineers in 1890. The Engineers in turn sponsored the final passage, after several failures, of a bill making central time the legal standard of Ohio in 1893.[82]

Ohio's hostility testifies to the lingering belief in an older conception of time — time linked to nature, to the sun, to a specific place and to that place alone. Accepting standard time as pure commonsense, we might judge their objections silly, backward, and provincial. But in fact most of the communities cited above held no grudge against commerce, no hostility to modern science and learning, no objection to the railroad as a clever device for moving people and goods. What each found puzzling, saddening, or infuriating was the assumption that time was arbitrary, changeable, susceptible to the whims of the railroads or defined by mere commercial expediency. Surely the world ran by higher priorities than railroad scheduling.

The Texas Circuit Court insisted in 1895, for example, that "the only standard recognized by the Courts is the meridian of the sun, and any arbitrary standard set up by persons in business will not be recognized." In even stronger terms, the Iowa Supreme Court declared in 1899: "we are not quite ready to concede that, for the mere convenience of these companies, nature's timepiece may be arbitrarily superseded." "The apparent daily revolution of the celestial body," the Court pointed out, "has from the remotest antiquity been employed as a measure of time." Local time, it insisted, "cannot be lightly set aside on the mere pretext that certain lines of business so demand."[83]

Between 1883 and 1915 standard time came to trial before the Supreme Courts of various states at least fifteen times. The Supreme Courts of Nebraska (1890) and Kentucky (1905), and the United States Court of Appeals (1907), all ruled that although standard time had been adopted in many places, local time still ruled the day unless the law said otherwise.[84] Standard time was also tried in Minnesota (1898), North Dakota (1905), Utah (1911), New York and California (1917). Though upheld in each case, the willingness to test standard time points to public awareness of alternatives, or outright hostility, to the system of time we now take for granted.[85]

These dissenters, to be sure, were decidedly in the minority — most communities and states adopted standard time readily and even enthusiastically, if the newspapers are to be believed. As an innovation, standard time added the weight of scientific endorsement and the power of the railroads to its eminent practicality and usefulness. In some cities, the reform passed almost unnoticed, or was treated in the matter-of-fact way befitting what almost seemed like "common sense." Standard time advanced with all the weight and momentum of industrial progress, and its opponents were clearly out of step with prevailing opinion in most published sources.

But neither should their objections be lightly dismissed. They objected not only to standard time itself, but to the new ideas about time and time usage it represented. Many raised questions of political, cultural, and religious authority that would find echoes in years to come. What principles, they asked, should govern our society? Most of the states which saw standard time brought to trial, for example—Georgia, Texas, Kentucky, Nebraska, Minnesota, North Dakota, California—were hotbeds of rural radicalism and Populist support. Iowa, whose Supreme Court ringingly denounced railroad time, gave birth to General James B. Weaver, while another standard time case originated in Iron Gate, Virginia, home of several Farmer's Alliance cooperatives and a strong Populist community.[86]

An analysis of Populist rhetoric, offered here only as the briefest suggestion, suggests that they rejected the kind of linear, "industrial" time implicit in the new standards. "All nations of history," declared the *National Economist* in 1889, "have had a rise, a period of brilliant prosperity, a decadence and a fall." Depicting history as a series of rises and declines obviously involves a cyclical understanding of time, and in the series on world history quoted, the Populist newspaper laid the blame for decline on neglecting agriculture.[87] "When Rome," another Populist rhetorician added, "suffered her noble citizens to 'crush out' the cultivators . . . her power began to wane." Similarly, "retrogression in American agriculture means national decay and inevitable ruin."[88]

Populist writings insisted on nature as the source of time and natural imperatives as guides to using it. "Land," lectured agrarian firebrand W. Scott Morgan, "is the gift of God." Morgan quoted the Bible to make a point about nature's model for time usage. "In the sweat of thy face shalt thou eat thy bread. This decree was uttered 6,000 years ago," he concluded. "It is as immutable as time itself." Farmers must cooperate, Morgan urged, and seek their model for cooperation "in the mechanisms of heaven, and in the animal, vegetable, and mineral kingdoms." Even the name of one of the earliest radical farmer's organizations—"the Wheel"— suggests cyclical time, with agriculture and natural law, rather than geared mechanisms, acting as a sort of social governor on the many wheels of industry.[89]

Recalling the almanacs and schoolbooks cited in Chapter One, these Populist writers objected not so much to machinery— or clocks, or industry, or the railroads—as to the ideas and assumptions about time and progress that governed them. None mentioned standard time explicitly, but each source framed its conceptions of history and good government in the circularity of "natural" time. The phrases they used to describe men's and women's relation to nature would echo again thirty years later in strong rural opposition to daylight saving.

If these cases suggest rural hostility to standard time and the culture that made it, the regulations cited below, drawn up by the General Time Convention in the wake of its success, hint at the new time's implications for industrial workers. These new guidelines called for "all conductors, engineers, train hands, road foremen," etc., to provide themselves with timepieces. Each was to take his watch to a jeweler or watchmaker, who would record the watch's serial number and issue a certificate of accuracy, to be renewed every three months. "All train men," the regulations continued, "should compare their time with standard clocks . . . regulate their watches therebye and register their names, and the time at which they register, in a book provided for that purpose," before every trip.[90]

These regulations, of course, seem eminently logical and practical. The point here is not just that railroads needed accurate time, but rather that in the wake of standard time such close attention to timekeeping was now common sense—standard clock time had become the regulating and organizing principle for industrial labor, and "common sense" about time now meant following an elaborate set of practices that reinforced the clock's authority. The virtue of ever greater "accuracy" was the path this time discipline traveled.

Compare these regulations, for example, with the Camden and Amboy's 1853 regulations cited in Chapter Two. Both emphasize the watch as timekeeper, and establish a set of routines for assuring its accuracy. And, not surprisingly, by 1884 the whole procedure has become more ritualized, more formal, and more elaborate. In 1853 workers had used company watches, and returned them at the end of the day. Now each employee was to provide himself with a watch and register his name daily next to the watch's serial number, to prove he had checked in with the standard clock. The daily registry testified to the accuracy of both worker and watch, while the standard clock served as the authority both man and machine were measured against.

The worker's regularity in signing the book, and in regulating his timepiece, defines the performance of his duty and thus his worth to the company. His accuracy, regularity, and dependability—certified by the registry—mirrors the watch's accuracy, regularity, and dependability as shown on the jeweler's certificate. But both exist in a subsidiary relation to their ultimate model, the standard clock. Standard time eliminated the kind of public—and private—conflicts that resulted when each man could make a legitimate claim for the authority of his own time. But it did so by merging men and women and their time-telling machines.

The *Indianapolis Journal,* for example, claimed that "there is something remarkable about the fascination which watches exert over their owners." Published two days before the establishment of standard time along with a favorable description of Allen's plan, the article insisted that "a man who prides himself on his watch identifies himself with it . . . the watch becomes a part of himself." The *Journal's* article, a discussion of time and timekeepers by a local professor, described the situation that standard time, and the railroad regulations, were supposed to prevent.

Among men who pride themselves on owning good time-pieces, the professor continued, "the infallibility of their watches is the firmest article of faith." But no watch keeps perfect time, and "meanwhile every man in his own breast doubts his watch. He gives an intellectual assent to the doctrine that his watch is infallible, but in the silence of the night he confesses to himself that perhaps it needs regulating." Spoofing the conventions of inner guilt and the nagging conscience, the article recalled Melville's *Pierre* and its intriguing connection of time with moral authority. Standard time solved the dilemma both Melville and the professor posed, by establishing an ultimate standard all were measured against.[91]

The article, and the regulations, testified to the questions of moral and political authority implied in railroad standard time—the exchange of local and individual autonomy for regulation by distant, even faceless (or rather, clock-faced) authority. Allen's innovation resolved public and private doubts about time and conduct by establishing a seemingly universal, arbitrary, and consensual standard. Standard time reaffirmed the priorities—commerce, trade, unity, the free exchange of goods and information above all—that now largely regulated the daily life of an industrial society.

Not all found that regulation comforting. On November 18, 1883, the day that standard time was implemented, the *Cincinnati*

Commercial Gazette ran a descriptive piece on that new phenomenon of industrial capitalism, the commuter. On any given day, the reporter found, the same pattern of "types" and classes among these "railroad regulars" repeated itself hourly at the local station. So regular were their movements that "the longer a man is a commuter, the more he grows to be a living time table—a model of cast-iron punctuality." The *Commercial Gazette* called the commuter's timetable "a hideous specter" that "sleeping or waking" haunted its victim. "It is upon his back," the article concluded, "and can not be shaken off."[92]

Though not writing about standard time per se, the reporter expressed some of the uneasiness the new understanding of time, and the newly developing relationship between men and women and their watches, provoked. Again, comments like these were strictly in the minority. But even those who celebrated standard time viewed its effects in ways we might find slightly uncomfortable. "Going down Broadway yesterday, I saw a man who stepped with odd precision," went an 1883 account in a New York paper, and "I called my companion's attention to this singularity." "He is running himself on an imaginary time-table," came the reply; "that is William F. Allen, the author of the new time system. He is an expert in Railway time-tables, and what you see in his gait and movements is a mannerism gained from his business."

Allen liked the account so much that he included it in his papers.[93] The description recalls the career of one of Allen's contemporaries, Frederick W. Taylor. In 1883 Taylor was measuring his stride to find a more efficient gait, and just beginning to apply the stopwatch to his fellow employees at Midvale Steel. The material cited here hints at some of the implications of standard time. Over the next twenty years the meaning of the new time the railroads fostered would reveal itself most clearly in the rise of scientific management and the growth of the American watch industry.

IV

Keep a Watch on Everybody

Trying to express the meaning of standard public time in 1883, the *New York Herald* reflected on its pervasive tendency to infiltrate ordinary life. Standard time, the *Herald* concluded, "goes beyond the public pursuits of men and enters into their private lives as part of themselves." The *Herald*'s comment implied that the railroad's innovation was more than a simple public convenience—that the change in the understanding of time marked, in fact, a reformation in the internal, private understanding of the self. Once individuals experienced time as a relationship between God and nature. Henceforth, under the railroad standards, men and women would measure themselves in relation to a publicly defined time based on synchronized clocks. The *Herald* knew standard time was an important change, but lacked words enough to explain why.[1]

A more pointed commentary came twelve years later from James O'Connell, president of the International Association of Machinists. Asked at a trade union meeting why labor might not form a trust like Standard Oil, O'Connell recalled the story of the Pennsylvania Irishman, who when told that his train left at eight o'clock "standard time," remarked irascibly, "Well, that settles it." "Settles what?" queried the train's conductor. "Why, the whole of it," replied the stock Irishman. "They'll be gittin' the wind next, they've got the time now." O'Connell's joke dramatized the exercise of power inherent in controlling and defining time. Standard time marked, as the *Herald* had realized, a major shift in the understanding and authority of time in everyday life. It turned time from an abstract quantity into a commodity, and more specifically—at least in the eyes of O'Connell's Irishman—turned time into a tool of corporate power. If Standard Oil could own time, then time itself became the laborer's antagonist.[2]

This is not to claim, again, that standard time *itself* caused any major change in people's lives. On the contrary, the new time zones usually amounted to just a small adjustment of watches and clocks. Most people met the innovation with praise and simply reset their watches, quickly forgetting all about railroad standard time, what it was and where it came from. In many cases, the railroads' 1883 innovation only formalized a regional trend. But standard time, in fact the whole concept of standardizing time, epitomized new ideas about time's nature, ideas that in turn contributed to a major reorganization of work, leisure, and the individual's relation to society. Standard time rebuilt the framework of time in everyday life, as the ideas about time that had inspired it left the railroad station and began moving into the work place and home.

Railroad time zones were only one manifestation of a whole range of new experiences of time. Electric lights altered the patterns of work and rest. Telephone lines brought the distant near and made conversation instantaneous. Clock and watch manufacturers doubled, tripled, even quadrupled their output, surrounding the time-conscious with a staggering variety of mechanical timekeepers in every size and shape imaginable. Each of these factors combined to create, in the decades after 1880, a generally height- ened sense of punctuality and urgency about time and clocks, a new and widespread formality in the experience of time in every- day work and life. But just as some Americans had consciously resisted standard time, so many Americans continued to resist the clock's authority over time in their daily lives.

Americans experienced one aspect of this reformation of time consciousness in a new emphasis on strict punctuality in work, in private life and at public events, theaters and concerts. Historian Lawrence Levine has recently documented the "bifurcation" of American culture in the nineteenth century, the conscious creation of a split between "highbrow" and "lowbrow" forms of art and entertainment. According to Levine the loose, relaxed, socially mixed quality of public events in antebellum America began to clash with a more circumspect, formal model of decorum advanced by the new arbiters of elite culture. As rules about public behavior changed, previously relaxed customs about starting times and times of arrival gave way to highly structured and closely enforced rules against lateness. "Unpunctuality," a Boston newspaper claimed in an 1884 example, "shows a relaxed morality in the musical community." Indifference to time stigmatized the late- comer as a moral leper and a social inferior.[3]

Ideally, this temporal self-discipline—the "standard time" of the work place—should have carried over into recreation, but too many people seemed to be ignoring it. "It is very singular," scolded a magazine article of 1892, "that in a country where railroads, with their inflexible time schedules, are such an important and universal feature of life, the habit of punctuality should not be more generally diffused." This editorial proposed using railroad standard time, so useful in organizing labor and business, as the model for ordering leisure. But while praising the ideals of clock time the editorial also remarked on a lingering resistance, in 1892, to the clock's authority in leisure hours. "Few people are habitually late to a train, to business, or to dinner, but many seem to consider it of little consequence," the author lamented, "to be from ten to thirty minutes behind time for every concert, lecture and theater." The same rigid schedules that the author found in work, business, and the home still lacked binding authority over public recreations.[4]

The latter decades of the nineteenth century saw, as this editorial reflects, both continued resistance to clock time and a constant process of negotiation and redefinition of time's role and meaning in daily life. In commercial life, the hegemony of standard time seemed fairly clear. For example, by the 1880s traditional schoolbook homilies on the evils of idle time were being fortified with strictures against tardiness of any kind. *McGuffey's Readers* from the antebellum decades had always stressed conserving time, but later revisions began emphasizing the urgent necessity of punctuality in meeting debts and following orders. "A railroad train was rushing along at almost lightning speed," one 1881 fifth-grade lesson began. "The conductor was late . . . but he hoped to pass the curve safely . . . in an instant there was a collision: a shriek, a shock, and fifty souls were in eternity, and all because an engineer was behind time." A commercial firm failed, the lesson continued, because its agent had been late in remitting payment, and an innocent man died because the watch of a messenger, who bore pardon, was five minutes late. "Napoleon died a prisoner at St. Helena," a final grand example claimed, "because one of his Marshals was behind time." "It is continually so in life," went the conclusion. "The best laid plans, the most important affairs, the fortunes of individuals, honor, happiness, life itself are daily sacrificed because somebody is 'behind time.' "[5]

If only Napoleon's marshals could have owned a set of *McGuffey's Readers*! *McGuffey*'s lesson embodies many of the new ideas about time typical of the late nineteenth century, in particular the *interconnectedness* of events. Time is expressed not in nature and seasonal tasks, but in terms of legal, industrial, and political obligations, conditions whose binding authority comes not from God, but from the interdependencies of a commercial economy. Standard railroad time, like the new emphasis on punctuality in schoolbooks and in public life, grew out of these new conditions, out of the same need to order and control financial events and human movements across space and time. It offered a uniform gauge for regularizing market relations, and *McGuffey*'s lesson reiterated the importance of keeping to the standard.

But resistance to clock authority continued—a third-grade *McGuffey's Reader* of the same year, for example, persisted in locating time in nature and the sun. One often reprinted story, "The Clock and the Sundial," related the humiliation of a boasting steeple clock, who had laughed at the humble and primitive sundial in the church yard below. The clock praised itself for telling time on cloudy days, and at night, but fell into embarrassed silence when the sun "broke forth from behind a cloud, and showed, by the sundial, that the clock was half an hour behind the right time." The moral cautioned against boastfulness, but the dial's triumph named the sun as the ultimate timekeeping authority. While daily life and business seemed to demand ever more attention to clocks and watches, tradition—and nostalgia—insisted on the wisdom and virtue of nature's example.[6]

"The chopping up of time into rigid periods," writer and editor Charles Dudley Warner wrote early in 1884, shortly after the invention of standard time, "is an invasion of individual freedom, and makes no allowances for differences in temperament and feeling." Warner was speaking about the difference between man-made and "natural" time, between relations of "sentiment" and relations of contract. He linked periodized, man-made time to creditors and their economic power—"they watch the manner of these artificial periods with interest," he punned, "in order to send in their bills and extort their profits." Warner mourned the apparent decline of "natural," precommercial time. The debtor, he claimed, "hates this artificial and vexatious arrangement of time" and "would like [time] to flow on unbroken like a river, peacefully, without dams, and without the constant apprehension of checks to his serenity."

Against the practical human divisions of clock and calendar Warner advanced natural metaphors for time, nostalgically yearning for an earlier period when life seemed freer. But he admitted that "in giving [time] a name and a certain space of duration we have made it an entity, and we cannot escape from the thralldom of this arbitrary succession." Warner's musings pointed out, like O'Connell's joke, that customary divisions of time also served as methods of control that privileged one economic class over another. Warner's time is not an indescribable "something" one exists in, a relationship between the individual and the natural world, but rather an assigned standard value, a number, that the less fortunate struggle frantically to keep even with. His comments, like the lessons from *McGuffey*, depict the constant tension between modern, arbitrary standard clock time and its preindustrial counterpart.[7]

In a similar vein, George M. Beard's pioneering 1881 study of *American Nervousness* partly blamed "clocks and watches" and the "necessity of punctuality" for the American middle class's generally strained nerves. For Beard, "clocks and watches" and the "necessity of punctuality" were nearly synonymous—they reinforced and mirrored each other. The very perfection of timekeeping instruments, he claimed, "compel[s] us to be on time, and excite[s] the habit of looking to see the exact moment." "A nervous man cannot take out his watch and look at it," he argued, "without affecting his pulse." Beard's watches and clocks "excited," "compelled," and increased the pulse—they laid claims on both body and spirit. Beard depicted a fierce and rigid social time discipline that demanded unhealthily intimate relations between people and their time-telling machines. If he shared Warner's distaste for the tyranny of standard public time, Beard perceptively located the problem in the peculiar—and fascinating—relationship between the watch and its owner.[8]

≡≡≡≡

Americans in the Gilded Age bought watches and clocks in huge numbers, in a vast range of designs, and for many different purposes. Between 1870 and the first decades of the twentieth century, the design of American clocks and watches, and the advertising and publicity they generated, consistently reiterated a few basic themes.[9]

Mechanical timekeepers stood for getting control—gaining power over the confusing and potentially hostile world of schedules, appointments, and standard time, or gaining power over unruly employees. But they also encouraged a strange merger of personal authority, of identity, between timekeeping machines and their owners. Clock time, machine time, savored unmistakably of discipline, surveillance, and control in both the design of the objects themselves and the discourse surrounding them. Yet the same objects that symbolized industrial and social discipline also empowered its victims. Standardized time marked no advance or decline in the extent of freedom or individuality. Instead, it represented a reconstruction of governing authority—both self-government and the organization of work and public life. What did it mean when ordinary people, engaged in the usual compromises required to earn a living, adopted machines as their models?

As the basis of time moved further away from nature and closer to the mechanical timekeeper, standardized clock time became a tool of education and industry, a sort of uniform that all laborers in the commercial army were required to put on at childhood and wear through life. And if standard time established the new battle dress of industrial competition, then the pocketwatch and regulating clock, mass-produced in unprecedented numbers after 1880, became its shield and insignia. But not everyone understood or used these weapons in the same way. Rural Americans insisted on tying time—and clocks—to nature. On the job, cheap pocketwatches offered an alternative to the boss's factory clock. And in the home especially, clocks became friendly expressions of individuality rather than grim sentries of industry.

Two themes predominated as standardized clock time spread: on the one hand, outright resistance to clock time, or the desire to control it, and on the other a peculiar concern with internalizing clock authority and finding one's niche in the new framework of standardized time. The two themes co-existed uneasily, reproducing a general uneasiness about time in society as a whole. One theme, followed in the industrial work place, led to time clocks, to Taylorism, and the ultimate establishment of machines as authorities for time and models for labor. The other led toward continued hostility to standardized time and factory discipline, to the idealization of leisure time, and to the bifurcation of public and private time in the home. Both themes revolved around the form of the mechanical timekeeper.

A number of innovations in timekeeping and time transmissions helped reinforce the authority of mechanical timekeepers after 1880. William Bond, and later Samuel Langley, had pioneered in transmitting commercial time signals, as Chapter Two related. Following Langley's profitable example, Leonard Waldo and the Western Union Company, among others, had established competing time systems of their own by 1883. Each offered a fixed authority for time—after 1883 the widely adopted railroad standard zones—and each claimed to lessen the confusions and anxiety allegedly stemming from idiosyncratic local times. But the proliferation of watches, time clocks, synchronized program clock systems, and time services over the next decades suggests instead that standardized time and time transmission only made people more anxious about time's passage.

Companies like Waldo's promoted synchronized clock systems. They linked "slave" or "controlled" clocks with a single, more accurate "master clock" located many miles away. An 1882 example, the "electrically controlled system" of P. H. Dudley, used signals from a local observatory to "control any number of clocks within telegraphic communication." When the Naval Observatory agreed in 1883 to telegraph standard railway time, it greatly increased the prestige of standard zones while creating the impression that its signals were the "official" standard of time for the nation.[10] Only Western Union offered subscribers the chance to receive standard time signals direct from the Naval Observatory. By 1893 they dominated the field through a partner corporation, the Self-Winding Clock Company.

"The Western Union Telegraph Company," promised the application for time service, "will furnish to its subscribers, once each working day, as correct Time Signals as it can obtain from the United States Naval Observatory Standard Time at Washington, D.C.," along with one or more clocks designed to receive the signals. "Said clocks," went the contract, "are the property of the Self-Winding Clock Company of New York, and are constructed and used under its patents." The relationship between the two companies was rather obscure, but the impressive effect of the partnership between the Naval Observatory and Western Union was not.[11] Regulated by electricity and far more accurate than daily life required, the company's clocks embodied time itself rather than merely depicting it.

The Self-Winding Clock Company's advertisements empha-

Tower and Pavilion of the Self-Winding Clock Company at the World's Columbian Exposition, 1893. The regulating clock, located in the pavilion, sent out signals governing the tower and the other clocks in the building. (From *Scientific American*, July 29, 1893. National Museum of American History, Smithsonian Institution.)

sized the Naval Observatory connection and the completely automatic nature of their service. Thanks to an electric motor, the aptly named clocks wound themselves, and in offices, factories, and homes were guaranteed to run for over one year without attention. At the World's Columbian Exposition of 1893 the company built an ornate, 150-foot clock tower demonstrating its wares, and boasted of regulating over two hundred smaller clocks from its central pavilion. A pamphlet accompanying the exhibit asked, "what is standard time?" The answer, written for the company by the naval officer in charge, was that standard time was the railroad meridian zones as measured by the Naval Observatory and registered on Self-Winding Clocks.[12]

"Standard time," in this sense, meant not "the railroad time zones" but rather Time itself. The establishment of a standard clock time, ticking away independently of local conditions, made standard clocks, rather than nature, the source of time—time becomes what the clock shows, instead of some abstract quantity the clock merely serves to represent. And as time became what the clock showed, the clock became the model for using time. The

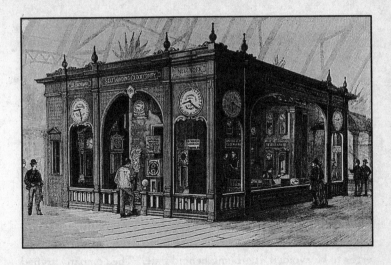

Self-Winding Clock Company, and its imitators, largely perfected the time transmission networks envisioned by Langley and others, making standard time service cheap, automatic, and accessible. There was no denying the virtue and even necessity of accurate standard time in this age, wrote *Scientific American* of the Columbian Fair exhibit. But previously, "only the most prosperous jewelers" could afford time signals. Western Union's service, the magazine enthused, "places standard time within the reach of everybody."[13]

But what would everybody do with standard time once they had it? Would they become automated "self-winders," like the clocks themselves? Previous systems of time transmission had suffered from a number of technological shortcomings. Electrical contact points, for example, frequently corroded or simply wore out. As manufacturers solved these and other problems a wide variety of related electrical clock systems entered the marketplace. Although not all sold Naval Observatory standard time — Western Union enjoyed a monopoly in this respect — each offered to run all the clocks in any office, factory, or school on a uniform standard. And each company also took great pains to associate

Time itself with their clocks. Like the Self-Winding Clock Company's, these systems did more than place standard time within the reach of everybody; they placed everybody within the reach of standard time.

The Electric Signal Clock Company's 1891 catalogue, for example, lauded the virtues of synchronized program clocks—basically clock/timers set to ring bells or trigger machines at certain intervals. Clock-bound bell ringers like these still administer the daily schedule in high schools everywhere, and most of us probably take them as one of the "givens" of modern life. But in 1891 their novelty demanded explanation. "If there is one virtue that should be cultivated more than any other by him who would succeed in life," the catalogue preached, "it is punctuality: if there is one error to be avoided, it is being behind time." The company argued that its best model, appropriately named the Autocrat, "gives military precision, and teaches practicality, promptness and precision" wherever adopted. A school, office, or factory installing this system, the pamphlet went on, "is not at the caprice of a forgetful bell ringer, nor anyone's watch, as the office clock is now the standard time for the plant." In other words, human authority is replaced by the clock—or rather, human authority merges with the clock's.

The Autocrat, the brochure claimed, did more than simply standardize time; it extended the supervisor's disciplinary reach beyond his office and throughout the building. It "revolutionizes the stragglers and behind-time people," went the explanation, since "there is no appeal from these signals—they are the voice of the principal speaking through the standard clock in his office." Thanks to the Autocrat, the principal's authority fused with the clock's. And according to the pamphlet's idealized description, the automatic pattern of master clock/controlled clock—the kind of regularity and precision that the Autocrat exemplified—would reproduce itself in employer/employee relations.

"It [the clock] will call 'time,' " the brochure proclaimed, "on the teacher who rides hobbies in public school work, who devotes fifty minutes to teaching Geography and ten minutes to Arithmetic."[14] The clock imposed mental discipline and regularity on teachers who lacked such discipline and whose enthusiasm threatened to run wild. A competitor, the Blodgett Signal Clock Company, claimed that "order, promptness and regularity are cardinal principles to impress on the minds of young people," and that "no better illustration of these principles than this clock can be secured in a school."[15] These program clocks exemplified and reproduced the supervisor's ideals of punctuality and discipline.

Thanks to synchronized clocks, "school officials, superintendents and principals will have the satisfaction of knowing that whether they are in the school building or absent from it . . . their schools are running on exact time," one brochure claimed.[16] The program clock in effect allowed the supervisor to be in two places at once. A testimonial letter from the principal of a Massachusetts high school affirmed that "no assistant in the school is superior to [the Blodgett clock] in promptness and faithfulness . . . I have no hesitation," the letter continued, "in recommending them to any school officer who is searching for a valuable (I might almost say indispensable) assistant in the school."[17]

A strange identification between the clock and its users permeates these phrases. The principal or supervisor's function seems in large part a mechanical one—imposing strict time discipline on recalcitrant employees and students. But these clocks not only imitated the supervisor's internal virtues, his discipline, his punctuality; they augmented them. The clock's mechanisms, like a sort of mirror that improved what it reflected, taught lessons in self-discipline, self-regulation, and self-improvement. "By its promptness and precision," went the claim, the Autocrat "has an invigorating effect on all about the school."

Why the word "invigorating"? Was the clock invigorating

because it took over certain tedious and mechanical tasks, loosing the creative impulse? Or did it invigorate in the same way that fear of the boss's wrath, and unemployment, invigorates? In theory, the Autocrat freed school personnel from the necessity of disciplining themselves about time — "it will save the teachers nervous force," went the argument, since "it prevents the necessity of watching the clock and teaching at the same time."[18] But installing a program clock would only make employees and students truly "self-winding," screwed to a tighter, more controlled level of tension about keeping to schedule. The very perfection of these clocks, as George Beard had suggested, only increased anxiety. By locating time in mechanisms, they made time discipline automatic and external — they made machines rather than people the keepers of time. But that very use, in schools and factories, made machine precision, and machine discipline, the prototype and model employees were forced to emulate in their daily work. The machine patterns human behavior.

Consider, for example, this illustration from the 1908 Self-Winding Clock Company catalogue. What did the prospective client see when he looked at this ad? Each room, except for those lower rooms reserved for management, is identical, transparent. Each room is empty of people, which implies that no matter who uses the room, no matter what they bring to the building, they will have an identical experience inside. Any group of people walking into this building would assume the pattern of discipline and regularity sent out by the master clock. Management too — despite the fireplaces, the books, the swivel chairs — the boss runs by the clock as well, even if he does have the power to set its automated alarms. The public building becomes a machine for instilling the clock's virtues.[19]

If synchronized clock systems like the Autocrat offered a pattern for the relations between employer and employee, between principal, staff, and students, then the invention of factory punch

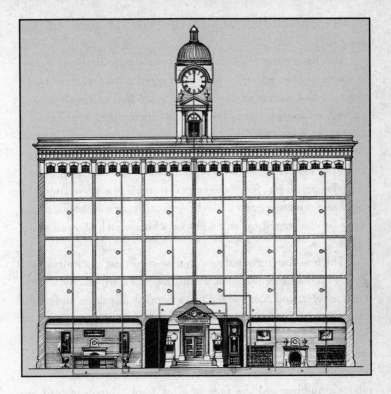

An example of master and slave, synchronized and programmed clock systems from the 1908 Catalogue of the Self-Winding Clock Company. (National Museum of American History, Smithsonian Institution.)

clocks—synchronized electric clocks with time-stamping mechanisms added—carried the tendency even further. Devices for recording laborers' time date back at least as far as Josiah Wedgwood's early factories, and the English inventor Charles Babbage designed a mechanical punch clock in 1844. The National Museum of American History has a mid-nineteenth-century weaver's lamp, a small, clear glass oil lamp marked with hourly gradations running from 8:00 a.m. to 6:00 p.m. As an inscription on the base proclaimed, the lamp provided "time and light" to laborers using it.[20]

But before the 1880s most factories simply stationed a man or woman at the entrance gate and logged a handwritten record of employees' names as they entered and left. The simple system worked reasonably well, but human timekeepers introduced a number of problems. They were slow—in larger firms anxious employees waited in line, losing wages, while the timekeeper ran down the list. They were inaccurate and subject to tampering—a timekeeper might forge the record to protect a tardy friend or penalize a punctual enemy. And the timekeeper's authority might easily be challenged by an employee whose watch showed a different time.

The first American patents on time-recording clocks followed shortly on the heels of standard time, from Willard Bundy, a New York jeweler, and Alexander Dey, a Scottish physician and mathematician. In about 1885, Bundy observed the clumsy methods then in use in local factories. Typically, each man or woman received a brass check with a number on it. On entry, he or she presented the check to a timekeeper, who then wrote the employee's number and the time in a log book. Bundy instead assigned each worker a brass key. Inserted in a special clock of Bundy's design, the key triggered a stamp that printed the number and time on a tape. Alexander Dey's system was basically similar, but his clock presented each employee with a large dial with numbers on it. The employee swung a pointer around to his or her number, then pushed a small rod, which printed time and number on a prepared sheet. Both men received patents in 1889 and went into business shortly after.[21]

Bundy boasted that under his system, "each employee is his own timekeeper," and that "favoritism and collusion are impossible." The clock's "infallible printed record" tended toward "encouraging promptness" while preserving the "order and discipline so necessary to the conduct of every well-regulated establishment." Bundy promised that his system benefited employees by protecting them from cheating, adding for the employer's benefit that the printed record of time surrendered itself only to those holding the special key that opened the clock.[22] By 1894 the first time card stamping system had appeared, and by 1907 nearly all the leading manufacturers in the industry had been bought up by the International Time Recording Company, later to be known as IBM.

The International Time Recording Company hit its stride in the second decade of the twentieth century. Its clocks, advertisements declared, allowed employers to record, organize, and control their own and their employees' use of time. "Evanescence," a 1916 catalogue proclaimed, "should be the watch word of every employer of labor." The company defined evanescence as "the state of being liable to vanish and escape possession," and its catalogue warned that "the avenues of 'escape' for valuable time are many." To those employers working against time's slippery escape the company stressed the accuracy, impartiality, and prestige of the printed record. "Should the fifty-hour week become a law," cautioned the company's New England sales manager in 1913, "you will doubtless require a more careful accounting of your labor costs than heretofore." With profit margins growing ever tighter, "minutes will count—a penciled record is not impressive, not sufficiently accurate." The company's clocks, on the other hand, would "save money, enforce discipline and add to the productive time." Even further, "the time recorder induces punctuality," a brochure proclaimed, "by impressing the value of time on each individual."[23]

In other words the company's clocks not only aided cost keeping, they improved the temporal morality of the plant as well. "There is nothing so fatal to the discipline of a plant nor so disastrous to its smooth and profitable working," went another pamphlet, "as to have a body of men irregular in appearance, who come late and go out at odd times." In a large factory the employees became faceless and indistinguishable—a typically double-edged aspect of industrialization. If the tardy colluded with human time-keepers, they might be able to escape between the cracks and go undetected. The time recorder helped management "to weed out these undesirables."[24]

One 1911 advertisement showed a line of thirteen men and women laborers approaching the time clock. Arrows reaching out from the clock pointed to four figures drawn in solid black. "Un-profitable employees," went the text, *"mean* to be on time . . . but experience shows that it is the same old story again and again with the tardy ones." Since "the habitual late-comers and early-leavers do not work so well" as the punctual, with the clock "you can sift out the efficient employees from the inefficient." You might never know which employees were inclined to be late and thus to work badly, the text implied, unless the clock revealed the inter-nal flaw—the failure to internalize clock discipline—in each per-son's character.[25]

The mechanical timekeeper, again, forms the model or pattern for labor, the authority for conduct. The time clock pointed its bony hands at any employee who managed, thanks to collusion or sheer anonymity in numbers, to conceal his or her deviation from the standard. As another pamphlet boldly declared, an In-ternational Time Recording Company clock formed "the con-necting link between employer and employee" where sheer size had severed it. The clock replaced the personal relations that regulated paternalistic factory management with electromechani-cal links.[26]

By 1910 that Holy Grail of factory management, efficiency, began appearing more frequently in company publications. A time system, clients learned, "affords an index to efficiency," unerringly flagging "efficiency failures" and "enabling the management to judge whether men and machinery are measuring up to the proper

*Advertisement, from a magazine of factory management, for the
International Time Recording Company (later IBM), 1911.*

standards of efficiency." The Progressive era felt a near obsession with the subject of efficiency. As one historian of the period has written, "efficient and good came closer to meaning the same thing in these years than in any other period in American history."[27] "Efficiency" meant self-mastery, conservation of time, and the elimination of wasteful activities, but in its progressive crusade armor it described social activism and missionary zeal as well as energy, modernity, cleverness. And in the history of American industry no single figure better embodies both the fetish of efficiency and its relation to time, timekeepers, and their authority than Frederick W. Taylor.

The "father of scientific management" sought efficiency above all—maximum yield with least possible waste. He claimed to have found the key to eliminating human obduracy and laziness, the peculiar and troublesome quirks or corruptions that always resulted wherever men and women worked for others. Adopted by the public at large after the turn of the century, thanks in part to Taylor's ready self-promotion, the word "efficiency" concentrated the familiar American tight-fisted conservation of time into a social rallying cry, a call to arms for the well-meaning and the educated.[28] Scientific management, or "efficiency engineering," as the systems of Taylor and his peers came to be known, became a pet reform among crusading Progressives—a veritable "efficiency craze." Based supposedly in scientific principles, it seemed to many capable of rationalizing a vast array of social evils by eliminating all sorts of generalized "inefficiency."

Scientific management also sought to rearrange the relations between individuals and society. In an efficient world, each person would work to his or her full potential, instead of according to some slipshod amalgam of tradition, chance, and misguided personal preference. Forget following your father's trade; forget cooking dinner as your mother did. Instead of struggling along gamely at a job that didn't really suit, you the efficient employee would have been scientifically evaluated and put to a task perfectly fitted to your unique physical and mental attributes. If that meant rote work on an assembly line, so be it—"efficiency engineering," undertaken by trained experts, would rearrange the priorities that governed life and labor. In its most benign guise scientific management promised a utopia where all individuals worked at their individual best. A frankly machinelike process, it directed itself at perfecting each individual's functioning within the larger system. Scientific management reached its zenith under Louis Brandeis's aegis in the Eastern Rate Case, and perhaps reached its nadir a few years later in 1915, when reaction against Taylor's work at the Watertown, Massachusetts Arsenal resulted in a prohibition of stopwatch time study in government contracts.[29]

Taylor came from a well-to-do and genteel Philadelphia family. He inherited an active and productive temperament from his Quaker mother, heiress to the sort of tireless, unstinting industry Max Weber identified in Benjamin Franklin. According to Taylor's laudatory official biography, "work, and drill, and discipline" characterized life in her house. A fairly peculiar boy, plagued by nightmares and revolted by disorder, Taylor would transfer his mother's organizational energies to the realm of machine production. At age eighteen he began working at the nearby Midvale Steel Company, owned by family friends, as an apprentice pattern maker and later a machinist.[30] At Midvale Steel his obsession with rationalizing ordinary habits and motions focused on the problem of time.

Historians often approach Taylor as if he represented some sharp and peculiarly "modern" break with preindustrial thinking about the organization of labor time. But in fact American history offers numerous antecedents to Taylor's rationalizing thrust. Catherine Beecher's 1841 *Treatise on Domestic Economy,* for example, equaled or excelled any of Taylor's writings in its emphasis on the moral necessity of squeezing the utmost value from the time at hand. And Beecher's model kitchen certainly anticipated Taylor's concern with minimizing needless motions and unproductive distractions. Like Beecher, Taylor worked in a familiar vocabulary, drawn from schoolbooks and children's stories, that most middle-class Protestants memorized and recited as the catechism of success. There was nothing new about his determination to extract the maximum yield from a given effort; in this Taylor simply chose one particularly vivid color from the palette of American thought. Taylor became a true innovator only when he approached the standardized time of factory production with his stopwatch, and attached the gospel of efficiency to the idea of standard time.[31]

Taylor maintained that on most jobs only a blend of custom, inherited experience, and calculated resistance to working harder determined a worker's rate per piece. By analyzing the motions that went into any given task with a stopwatch, Taylor tried to find the minimum time in which the work might be done—to find an unvarying, uniform, and supremely efficient "standard time" for any job. Using stopwatches, Taylor timed the component movements of the job or set of jobs making up an industrial process. He separated "waste" motions—like scratching the head, hitching up the pants, or talking—from the movements he judged productive, and by combining these productive motions established the correct and scientific "standard time" for each worker's task.

Applying the same techniques to the factory as a whole, he reorganized the flow of information and materials, regulating all under the authority of a central planning office. Take, for example, the manufacture of machine parts. Taylor would begin by observing work processes—say, perhaps, unloading raw ore from freight cars. Did the workers make any "extra" motions? Did they stoop when they didn't have to? Were they using the right tools, at the proper speeds? Which were the actually *productive* motions, and which were simply bad habits?

After establishing a minimum time for each stage of the process from shoveling coal for fuel, to molding, to casting, to finish work on a lathe, Taylor and his disciples linked the component jobs together in a routing procedure regulated by a master clock in a central planning office. At the start and finish of each step in the production process, workers went to a secondary "slave" clock which stamped a card. The cards then went to the central planning office, where the "time clerk" checked them against the official standard. The planning office would ensure that each part of the process took precisely the correct time—fifteen minutes exactly for molding, ten for rough machining, and so on, till the piece had passed through the factory and was finished.[32]

Though the stopwatch, then and now, tends to symbolize Taylor and his work, in fact most of the supposedly "scientific" principles behind stopwatch study amounted to little more than guesswork, as angry laborers managed to point out. The genuine effectiveness of the Taylor system, in any factory or plant, depended at its core on a total managerial reorganization using synchronized master and slave clocks like those described above. Stopwatches might be used to determine the standard rate for the various jobs that went into making a part, but only the factorywide management system, directed by time clock from the planning office, ensured the efficiency and harmony of the whole operation.

Taylor was primarily a "system builder." His most lasting innovation was not stopwatch study but the reorganization of plant management. Though Taylor rarely mentioned it, the importance of time-recording systems to his entire process is unmistakable. By 1900 time-recording systems were undoubtedly simply taken for granted, and emphasizing their importance would have diminished Taylor's pretensions to innovation. But they were nevertheless essential to the process. In one of his first actions at the Watertown Arsenal, Taylor's assistant, Carl Barth, installed a new system of specialized time cards, time-stamping machines, and secondary clocks regulated electrically by a master clock in the planning room. In the ideal factory, Taylor insisted, each worker became "one of a train of gear wheels" regulated, like the plant itself, by the standard clock in the planning office and its "slave clocks" throughout the plant.[33]

Taylor's system evolved from the intellectual and technical innovations—like synchronized program clocks—that accompanied standard time zones. Taylor first conceived of time study either in 1881 or 1883–84, depending on either the definition of time study or Taylor's varied accounts.[34] The point is not the specific date but the fact that Taylor began his experiments precisely as the movement for national standard time peaked. Like the astronomers and scientists who advocated standard time, like William Allen and the railroad managers, or Bundy and Dey, inventors of the punch clock, Taylor hoped to crystallize "time" in mechanical timekeepers whose authority resisted dispute.

"Evanescence," the International Time Recording Company had suggested, should be the watchword in any modern business. Railroad standard zones ideally regularized time's flow, helping eliminate the local variations that had made time such a troublesome, "evanescent" industrial commodity. Using the standard time fixed in regulating master clocks, Taylor took that now uniform commodity and broke it into pieces. Taylor's stopwatch studies pinned evanescent time down for deliberate and close scrutiny.

It was the stopwatch, not the standard time clock, that came to stand for Taylorism and galvanize its enemies, and rightly so. Taylor's ideas came under close public scrutiny when his attempt to reorganize the government's Watertown, Massachusetts, Arsenal, which manufactured gun carriages and other weapon systems, ended in a strike and a congressional hearing. As historian

Daniel Rodgers pointed out, even a quick perusal of the testimony in the Watertown Arsenal hearings dramatizes the workers' intense hostility to stopwatch study. "I don't object to their finding out how long [a job] takes," testified machinist Orrin Cheney, "but I do object to their standing over me with a stop watch as if I was a race horse or an automobile." Another testified that he "did not believe in putting the stop watch on any man." Machinists and molders at Watertown repeatedly objected to having a man "stand over them with a stop watch."[35]

In the hands of a hostile observer, the stopwatch dramatized the worst aspects of mechanization in the factory. Even Watertown's most militant strikers assented with the principle of timing labor to determine a proper rate, and nearly all declared their allegiance to an honestly determined day's work. They were well accustomed to punching time clocks, and voiced no objection to them. Indeed, master and slave clocks with auxiliary time stamps had been installed at Watertown well before Taylor's arrival, with no complaint. What they resented most was the stopwatch's tendency to lend the phantasm of scientific impartiality to yet another form of sweated labor.[36]

Taylor knew the anger a stopwatch might provoke. He even acknowledged the "occasional" utility of hiding the stopwatch in the hollowed-out body of an ordinary book while timing went on.[37] The Watertown molders knew more than they cared to about stopwatch time study and were already grumbling about its use in other shops at the Arsenal. When the molders learned that one of Taylor's assistants planned on timing their work the next day, they secretly timed Taylor's timer with an ordinary pocketwatch. Taylor's man, using his stopwatch to time only the *productive* motions while ignoring "waste" motions, arrived at an ideal time of twenty-four minutes. Perkins, the molder, used an ordinary watch to time the entire job from start to finish. Making no attempt to separate "productive" from "unproductive" motions, he arrived at a total time of fifty minutes. The discrepancy in times outraged the molders: they struck the next day.[38] The resulting massive congressional inquiry into the utility of scientific management is their gift to history.

Both sides of the dispute rested their arguments on the watch's evidence. Taylor's man, the molders pointed out, lacked even the slightest experience with the molder's craft—he based his judgments of productive time on apparently arbitrary and mysterious criteria. The molders, on the other hand, derived their figure from the ordinary, common-sense time or duration of everyday experience. A man started at nine and finished at nine-fifty—what could be more straightforward? "Do you think you could get a practical record of how long it takes to do a piece of work," a Watertown machinist was asked, "without having a stopwatch?" "Yes," Olaf Nelson replied; "I think my watch will do, the one I carry in my pocket is close enough for all practical purposes." Nelson's ordinary watch measured the immutable flow of Time. But the stopwatch fractured Time to suit the boss's agenda. Under scientific management, as the molders' union representative put it, "the stopwatch is equivalent to a whip."[39] As a whip cut the air and the skin to discipline labor, Taylor's stopwatch cut and sliced Time itself to impose the machine logic of scientific management on human movements. With the stopwatch Taylor's assistant literally took time into his own hands, and wrenched it around and reordered it to fit Taylor's vision of time efficiently used.

Perkins's pocketwatch galvanized the opposition by giving the molders an alternative authority for time. The watch made them see Taylor's system, with considerable accuracy, as an elaborate deception. They acted to protect their control over the terms and pace of their labor, to preserve their authority as experts and masters of their craft. But they waged their battle over time and its definition. Whole time was their time, stopwatch time the re-structured time of machine-derived logic and the imperatives of efficiency. At the Arsenal, the pocketwatch became their best weapon.

At Watertown Arsenal the pocketwatch empowered the workers—it allowed them to counter Taylor in a rough semblance of his own language. But as E. P. Thompson pointed out in his exploration of time in industrializing England, their resistance also constituted an accommodation, a concession to the terms of the new factory system. Industrial capitalism triumphed, Thompson concluded, when workers accepted the clocklike regularity of the factory work day and began working to shorten it—when they "accepted the categories of their employers and learned to fight back within them."[40] In resisting Taylor's stopwatch, the Watertown strikers only affirmed the factory system and clock time.

There is no doubting either the triumph of the "second factory system" or the importance of Taylorite notions to its success. Nor is there much room for doubting that workers at all levels were drawn both willingly and unwillingly into its embrace. The Watertown Arsenal strike and the hearings it sparked brought the philosophy of the new factory system before the public for the first time. Taylor's ideas merited investigation because, as Congress correctly realized, the "management revolution" intimated a larger social and cultural reorganization of time and life patterned on clock efficiency. When schoolmasters installed synchronized clock systems, they reorganized education on machine models. When the acolytes of high culture demanded greater punctuality from their audiences, they sought to manage the audience's time more efficiently. To some, the emphasis on conserving time even seemed to be corroding the contemplative values of high culture itself. Henry Cabot Lodge, for example, deplored the quest for efficiency and greater speed for its pervasive effects on culture. "It has deteriorated style, it has deteriorated literature, it has deteriorated art," he complained on the House floor. "It is deteriorating manufacture."[41]

<hr>

Late nineteenth-century Americans, like the Watertown strikers, witnessed a reformation in public time—a greater emphasis on accuracy, on punctuality; a greater level of anxiety about watching the clock. And along with this reformation came a veritable explosion of suddenly affordable timekeepers. Mass-produced in an enormous variety of sizes, colors, shapes, and grades, the promotion and design of these watches and clocks adds depth, nuance, and even contrast to Thompson's portrait of militant defeat. In their variety and diversity these mechanical timekeepers reflect vast differences in the understanding and apprehension of time. And in their popularity, their sheer abundance, clocks and watches, the roles they played and the literature they generated, contained one of the central problems of industrialization—what effect would mechanical inventions have on their inventors?

A Yankee mechanic and watchmaker, Aaron Dennison, founded the first American watch factory in 1850. Known early on by shaky finances and a succession of names, Dennison's Waltham Watch Company flourished during the Civil War. Waltham's remarkably inexpensive "William Ellery" model brought watches within the common soldier's reach for the first time, albeit at some sacrifice. Dubbed "the soldier's watch," the Ellery accounted for nearly half the company's sales by 1865.[42] Inspired by Waltham, a number of new companies entered the field, improving manufacturing techniques and driving down prices in the process. By 1876, American mass-production innovations allowed Waltham and Elgin, the two largest companies, to produce over 200,000 watches a year. The combination of high quality and low price that Americans attained threw the Swiss, then the world leaders in watchmaking, into despair. In 1876 a Waltham watch might cost three times less than a Swiss watch of comparable grade, while the quality of Waltham's workmanship, one Swiss reported, was simply "incredible." "One would not find one such watch," a representative of the Swiss industry declared, "among fifty thousand of our manufacture." Prices declined as the century progressed. By the 1880s reasonably good American watches sold for as little as three to five dollars, and in 1898 the Ingersoll Company offered its popular dollar watch, dubbed "the watch that made the dollar famous."[43]

As David Landes points out, American watchmakers led the field not only in manufacturing but in advertising and promotion as well. "The outpouring of timepieces," Landes writes, "was accompanied by a comparable flood of publicity." As early as 1859, Waltham boasted with typical hyperbole that its watches rivaled the sun's accuracy. The Elgin Company went Waltham one better—its official trademark showed Father Time in silhouette, abandoning his hourglass and rushing off with a new Elgin watch. In Elgin's drawing Father Time looked like the venerable figure on *The Old Farmer's Almanac,* but he raised a small cloud of dust as he bustled about on his busy errands.[44]

The demand for watches mirrored the changes in public time and timekeeping. Elgin's Father Time, like America, was making progress, innovating, and speeding up the pace of his labor, and American watches announced their owner's membership in the fraternity of success and progress. In Horatio Alger stories, for example, the hero nearly always gets a watch upon his entry into the middle class. In Alger's stories, "the new watch marks the hero's attainment of a most elevated status, and is a symbol of his punctuality and respect for time . . . Alger makes much of the scene in which his hero receives from his patron a gold pocket watch, suitably engraved."[45] Alger's watches grant a degree of power to their owners — in *Struggling Upward, Or Luke Larkin's Luck*, Luke's watch is like the lance, shield, and armor conferred on some medieval knight. It expresses his identity as a good servant of industry while helping him do his job.

Henry Ford, tinkering with his neighbor's watches in 1883, claimed later to have realized that a watch might be manufactured for as little as thirty cents. But he never followed up on that insight, assuming that "watches were not universal necessities, and therefore people generally would not buy them." As he freely admitted later, Ford was badly wrong. In an age of increased attention to time, watches helped their owners cope, like Alger's heroes or the Watertown strikers, with appointments and schedules. Americans were so anxious to own watches that poor but watch-hungry citizens formed "watch clubs," to which each member contributed a certain amount per week. At the end of a week the members drew straws, the winner using their collective funds to buy a prized timekeeper.[46]

Then even more than now, owning a watch made everyday tasks easier, since consistently accurate public clocks were still hard to find. Accuracy, not surprisingly, was the single most consistent theme in watch advertising. But the expanding watch market drew on more than the simple need to tell time. A watch made a definite statement about its owner's relation to society as a whole. In *Struggling Upward*, Alger reiterated the hierarchy of watch ownership: gold was better than silver, silver better than gold plate, gold plate better than base metal. Different grades of watch ideally marked a boy's progress through life. The watch in this case served as more than a simple timekeeper — it established the owner's identity, symbolized his or her social status.

Echoing the relationship between the school supervisor and the supervising clock, the most striking theme in watch advertising stressed the close and peculiar identification of the watch with its owner. American watch manufacturers early on directed a vast arsenal of psychological appeals toward this curious relationship between the two. The Waterbury Watch Company especially loosed a flood of pamphlets and booklets aimed at children, offering insipid adaptations of nursery rhymes and songs—"Little Miss Muffet, sat on a tuffet, winding her watch so merry," bleated one.[47] "Children," asked another, "did you ever see a Waterbury watch? In many respects it is like you." How, you might reasonably ask, is a child like a watch? "Its face is always bright and shining," answered Waterbury, and "its hands are never idle, except when run down, which is like going to sleep."[48]

The simple identification of biological mechanisms with machines was an old idea, dating back hundreds of years and still common in the 1880s. One 1887 ad for patent medicine, for example, compared "the human system" to "a kind of machine. If one part goes seriously out of order the whole goes wrong."[49] What Waterbury suggested was something more; its watches taught their owners lessons in social and personal regularity. "It is a good idea to wind [the watch] often," the ad cited above continued, "just as it is a good plan to wash your hands often." A Waterbury "teaches a useful lesson, by always being regular and trustworthy. It is never late, like some children are at school. It helps children to be punctual and prompt."[50]

The identification between the watch and its owner was more than just some advertising firm's clever idea. It repeated itself in the mechanical peculiarities and design of the watches themselves. Telling time with a nineteenth-century pocketwatch is a very different experience than simply glancing at your wrist. Nineteenth-century watches have a certain weight and size, a substantial heft. A heavy gold watch registered its value, and its owner's success in life, conspicuously, like a big belly or an extravagant meal. In its size and mass and in the showy gestures involved in taking it from the pocket, opening the case, and announcing the time, the watch pronounced its owner's status. And like gold or silver coin, a gold or silver watch transcended the fluctuations of an unstable economy by freezing transitory economic value into solid form. Tucked into the bulging vest of newspaper-cartoon capitalists, the gold watch helped symbolize the Gilded Age plutocrat. Ironically, like most watch owners, that plutocrat was literally chained, a kind of slave, to his watch. Yet ornaments worn on the watch chain—lodge emblems, cigar cutters, small golden tokens of achievement—testified to the owner's progress through life, and connected his progress to punctuality and respect for time.

But most commonly watches recalled the present, and the modern individual's need to synchronize him- or herself with the passage of public time. The mechanical design of watches in particular mirrored this association. Pocketwatches were originally wound with small keys. For example, the Phi Beta Kappa key, symbol of academic achievement, was originally modeled after a watch key. American companies pioneered in the mass production of stem winders, wound instead by a knob at the top of the watch. Their innovation entered the vernacular as a synonym for the new, the progressive, and the clever. The phrase, "He's a real stem-winder," described a young man on the rise. It also implied, like the watch itself which required no key, that the young man came out of the box prepared for industrial employment and ready for anyone's use. The stem winder reflected standardization of jobs and of employee behavior.[51]

Even more suggestive is the peculiar fact that spring-driven watches, if wound at precisely the same time every day, become more accurate. "The winding of a watch is a matter requiring special attention," one instructive text proclaimed in 1883. "Some special hour should be selected for it." "The man, woman, boy or girl who will not take the trouble to wind a watch regularly," the author concluded, "is an undeserving and unreliable person who cannot be depended upon, and is not worthy of the cheapest watch in existence." Winding the watch at the same time every day minimized variations in the mainspring's force and thus in rate. Books and articles on timekeeping mentioned this point frequently, most notably the United States Bureau of Standards, which by 1910 had become the sine qua non of things accurate. In 1914 the Bureau suggested that if a watch owner passed some source of standard time every day on his way to work, he should wind his watch by that clock at precisely the same time every day. Standard time regulates the man, and the man's regularity regulates the watch, which in turn regulates the man again.[52]

"A watch seems almost a part of a man," one 1887 advertisement claimed. "It is next to a living creature in its steady work and usefulness." Pocketwatches were probably the only example of a clearly industrial machine carried close to the body. Watches, one journalist claimed, were "the things most alive and human in the entire range of [man's] handiwork."[53] Waterbury published a monthly magazine extolling its products to jewelers and retailers. One 1888 illustration, titled "Evolution," depicted a short, round man who bought a watch "that so pleased him that he could not stop looking at it, and from constant attention he underwent the following series of changes." Six drawings followed; in each the man came to resemble the pocketwatch more and more—his head becoming the stem and knob that wound the watch, his body growing rounder. Finally the "Evolution" ended when the little man literally became the object of his fascination—a Waterbury "series E." stem winder.[54]

The implications of this advertisement seem more than a little unsettling—Waterbury's version of "Evolution" leads toward an inevitable fusion of man and machine. In fact for more tradition-minded Americans, this was precisely the problem with Darwin's theory. Understood through the "social Darwinism" of Herbert Spencer and William Graham Sumner, evolution seemed like a cold, heartless, mechanical process, a relentless industrial competition that crushed human values beneath it. As we will see in

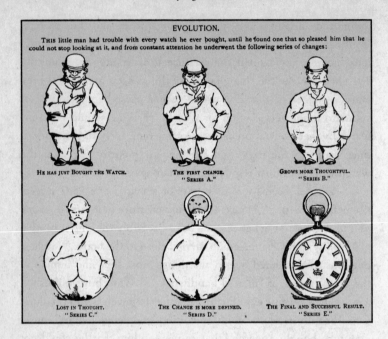

1888 illustration from **The Waterbury,** *an advertising almanac published from about 1887 to 1895.*

the later chapters, efficient machines became analogous to the "survival of the fittest" in human society. Would men and women have to become like machines—amoral, running with mechanical regularity—in order to survive industrial society?

In antebellum America, time had been resolutely linked to social discipline and control. Politicians warned against idle hands, ministers urged their parishioners to "redeem the time," schoolbook homilies reminded children of the busy bee and industrious ant. But the crucial difference between the earlier period and the Gilded Age lay in the source of time. When time came from nature, men and women were part of that relationship—they lived *in* nature and *in* time, since the two were the same. Contemplating nature meant, in great part, contemplating the vast span of time nature encompassed, and men and women's place in it. In Waterbury's advertisement, contemplating the watch led to the same sort of merger. But in the late nineteenth century, merging identity with the watch seemed to mean absorption into the public system of industrial time discipline.

The ideal of a disciplined and regulated industrial work force, which runs throughout these advertisements, is simply inescapable. The cover of a similar pamphlet, published by Waterbury in 1887, depicted a board fence, plastered over with handbills bearing the names of several famous anarchists and some of their associated themes. "WORKINGMEN ATTENTION," the pamphlet declared. "In these times it is necessary to KEEP A WATCH on everybody." "Your time is at hand," it continued. "Eternal vigilance should be your watch-word. Keep a watch on everybody. On the statesmen. On the capitalists. On the clergy. On the Police. On yourselves." The pamphlet used the materials at hand in 1887—standardized public time, clock-time discipline in the factory, fear of hidden and overt class resentments—to make a joke out of the connection between mechanical *watches* and political, industrial, and social surveillance and control.

The illustration on the last page showed a solitary workingman, walking the city at night, being accosted from behind by a uniformed policeman with a lamp in hand. Behind them stands the same collage of anarchists' posters and slogans that made up the cover. The officer claps his hand on the workman's shoulder, but the man exclaims, "No need to keep a watch on me, Mr. Cop, for I already have the best watch in the world—THE WATERBURY."[55] The pamphlet blurred the distinction between owning a watch, watching, and being watched. There is no need to observe the watch owner, it asserted, because in buying the timekeeper he has announced his membership in the class that needs no watching—that group which is already safely under

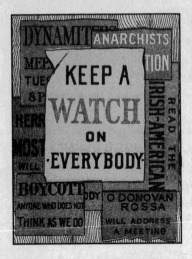

Front cover, back cover, and last page of a pamphlet ad for the Waterbury Watch Company issued in 1887. (National Museum of American History, Smithsonian Institution.)

surveillance and control because it accepts the mechanical time-keeper's authority.

Some advertisements, like that cited above, betray considerable uneasiness about the close relations between machines and the self. One of Elgin's 1875 advertisements depicted two boys and their teacher in an old-fashioned schoolhouse. The first boy, captioned "Waiting to be adjusted," stands by as the second, labeled "Adjusted to position," receives an apparently well-deserved caning.[56] The phrase "Adjusted to position" described a watch that had been certified to run accurately in up to seven positions—upside down, face down, on various edges. True, this was obviously a joke. But all jokes must have some correspondence with everyday life. Elgin's joke, like the Waterbury pamphlet, equated the watch's role as a "regulating" device with punishing authority.

These ads promoted the internalization of machine time, advancing watches as models for behavior, a sort of mechanical prosthetic conscience. It might be suggested that identifying consumers with the products they buy is an old advertising trick, and indeed represents nothing either new or startling. But in 1890 the identification between consumer and goods was a new idea—watch advertising may even have pioneered in this approach. More importantly, there is something different about watches, because of their close connection to the idea of time.

Louis Mumford once claimed that "the clock, not the steam engine, is the key machine of the modern industrial age." Mumford referred to both clock technology's contribution to machine design, and the role of the clock in regulating factories and timing labor.[57] To Mumford's claim could be added the mechanical timekeeper's role as an expressly *political* device, as a model for both self-government and the regulation of social groups. Where natural examples had once provided both the evidence of time's passage and a guide for its use, now in the environment of American industry, machines modeled behavior and their presence assured order and self-government. In the wake of standard time, and the rise of synchronized time clock systems, watches became little personal primers, little examples of the nature of good conduct and the social mechanisms that enforced it. To "keep a watch" on employees, and separate deceptive, unreliable workers from more regular ones, employers began relying on punch clocks shortly after the invention of standard time. To attest to their membership in the reliable class, American consumers wore their vest-pocket watches close to the heart.

It might be a mistake to claim too much for these advertisements—the fact that one makes a joke of surveillance indicates that surveillance, perhaps, was less threatening than a gloomy historian might be tempted to make it. "Keeping a watch on everybody," after all, was what led the Watertown Arsenal's molders to their successful strike. In many cases watches were posed as a means of *gaining* control rather than *being* controlled. Watch manufacturers offered their wares as the only way of coping with industrial time's imperatives. "When you have no watch," Waterbury claimed, "then time becomes aggressive and the moments gain the upper hand." Waterbury's products helped stave off such deplorable lapses in "your frantic daily struggles with time." One 1889 Waterbury promotion, for example, showed a woman and her son asleep and dreaming in different rooms. The boy dreams of "a watch as good as Papa's," the mother of "a lady's watch that will really keep time."[58] The illustration connected time and watches to power, especially adult male power. Owning a watch, the ad implied, made both woman and child more like a man, more like those in power.

Early in 1888 Waterbury revised the familiar tale, quoted from *McGuffey's Readers* early in this chapter, of the clock and the sundial. In Waterbury's version the boastful clock and the modest dial still show different times, but now neither prevails. Rewritten, Waterbury's moral advised: "Carry a Waterbury watch, which is never wrong, and be independent of either dial or town clock." The advertisement promised to give the Waterbury owner autonomy and personal authority over the time. In a busy world, Elgin warned in 1909, "time may be gained by the miracles of rail and wire, only to be lost by the inaccuracy of a watch." Hand in hand with modern time savers—"the limited, the telegraph and the telephone"—went the modern timekeeper, the Elgin watch. An Elgin synchronized its owner with the confusing, interrelated world of rapid communication and travel.[59]

The demand for watches brought prices steadily downward after the Civil War. By the 1880s, cheaper pocketwatches were regularly given away with suits of clothing, or awarded for selling magazine subscriptions. Newspapers abounded with such promotions. An 1883 issue of the *Cleveland Gazette,* for example, offered "this handsome stem-winding Waterbury watch and fifty dollars cash!" to "anyone who will sell fifty copies of The Life and Times of Frederick Douglass."[60] But the most intensive sales of watches took place in the mail-order catalogues that sprang up in the 1880s. Sears, Roebuck began in 1886, when Richard Sears, an enterprising railroad station agent in rural Minnesota, wound up with an unwanted shipment of pocketwatches. Sears sold the watches, at slightly above cost, to other agents along the line, who in turn sold them locally for significantly less than a jeweler's markup. He soon began turning this profitable trick on a regular basis, and later that year quit the railroad to form the R. W. Sears Watch Company. Within a year he was writing advertising copy for the burgeoning business.[61]

Like his main competitor, Montgomery Ward and Company, Sears aimed mostly at the rural market. Ward's identified itself as "the original Grange supply house," while Sears claimed to democratize watch ownership. "The boys, the girls, the rich, and the poor," Sears promised, "can all carry watches at the prices we ask." Both emphasized their opposition to the middleman and the parasitic trusts that leached extra profit from the isolated farmer. In one of its several incarnations, Sears advertised frequently in the Populist's weekly *National Economist*.[62] In 1897 Sears prefaced thirty-nine pages of catalogue watch ads with a warning about the local jeweler. "Step into his store at any time of year," the catalogue jeered, "and you will find him gossiping with a neighbor merchant, reading the newspaper, or both he and his clerks are cleaning up and polishing some of the old shop-worn goods . . . trying to make them look like new."[63] Sears depicted the merchant as an idler and a fraud. Producing nothing, he tries to put a shiny face on his own sloth. Even more significantly, Sears attacked the merchant's temporal world—"Step into his store," went the warning, "at any time of year"—the jeweler's trade was unaffected by the seasonal cycles of natural time.

The sheer abundance of watches in mail-order catalogues argues for the importance of time telling among rural Americans. And indeed this should come as no surprise, given the near obsession with time measurement routinely displayed in that most stereotypically rural document, *The Farmer's Almanac.* But the fact that farmers wanted watches does not necessarily indicate an endorsement of industrial time. Chapter Three discussed rural opposition to standard time, and hinted at the connection between the cyclical time of nature and Populist rhetoric about the moral virtue of agriculture. The design and marketing of these watches suggests continued rural hostility and resistance to standard time and the understanding of time it represented.

Early on, Sears realized the importance of showing the merchandise, and copious illustrations marked his catalogues. His 1889 catalogue, eighty pages long, devoted seventy-three pages to men's and women's watches and watch chains. In Sears's day most better pocketwatches were sold as "movements" alone—the basic mechanical works with a dial and hands. After choosing a movement, buyers then picked out a case from a wide variety of sizes, materials, and styles. The vast array of case styles parading across the catalogue pages of Sears and Montgomery Ward overwhelmingly connected time to rural life. Of the 228 cases illustrated in both catalogues in 1895 and 1897, the majority depicted some rustic bucolic scene—a rural village, a country church, a farmhouse; a horse, a hunting dog, flowers and leaves, or a pair of birds on a fence. Only six depicted an even vaguely urban or industrial scene, and all six were of locomotives. True, the engravings often had a "machined" appearance, and combined leaves and flowers with "engine turned" patterns. But the preference for familiar rural archetypes and symbols is unmistakable.[64]

The most popular style by far, the "hunting case," had a hinged cover over the dial, engraved on both front and back. To look at the time, the owner had to take out the watch, turn it face up, and open the cover. At first glance he or she saw a reassuring symbol of rural values, a representation of country life. The owner might use the watch to cook a meal, or to catch a train, but no matter what time the dial showed, the engraving on the case amounted to a ritual identification of time with nature. These case designs linked commercial or industrial time, in the form of the watch, to its natural origins. They reflected a basic ambivalence about time's new origins in machines.

Lavish decoration also made these watches like jewelry, like ornaments rather than essentially utilitarian machines. Why, after all, would a farmer living away from markets or towns even need a watch? These watches were less timekeepers than declarations of self. They linked their owners not to the specific work they did—there are no depictions of haying, for example—but to an idealized vision of the world they lived in. For women this was especially true. Women's watches had always tended to be smaller and more highly ornamented, more decorative than functional. Montgomery Ward sold the "Lady Grange" model, for example, with "a beautiful solid gold Victoria chain, and a handsome oak case, silver trimmed, with lock and key." More jewelry than time-keeper, watches like this made clock time into a luxury, a bauble adorning the successful farm wife.[65] "Lady Grange" may have been at least partly identified by her ownership of the watch, but the watch signifies her status more than her work—such a watch probably went along to social occasions or shopping trips rather than timing supper in the kitchen. It symbolized not work, but status enough for leisure.

Ward's advertisement—and indeed the rise of mail-order consumer goods in general—brings us to the subject of time in the home, and the difference, if any, between men's and women's experience of time. In many ways men's and women's understanding and interpretation of time seems nearly identical. Historians of American women have noticed the overwhelming concern with time saving and efficiency that runs throughout the historical literature of domestic feminism, especially the importance of spatial reorganization in housework. Women involved in American utopian and communitarian movements, like the Shakers or the Oneida collective, systematically rearranged domestic work spaces to make household labor both easier and more productive. Catherine Beecher's immensely popular books on domestic economy elevated this spatial rationalization and temporal efficiency to a creed. Like the communitarian feminists who preceded her, and like her advice-giving peers, Beecher shared the general American preoccupation with saving time. Throughout her career she reiterated the fundamental morality of tireless labor at home. "We have no right," she insisted, "to waste a single hour."[66] In this sense, the strong emphasis on saving and conserving time, women's experience of time at home probably differed little if at all from men's.

Almost without exception, nineteenth- and twentieth-century manuals of home management and housekeeping rang with disdain for idleness and waste. Open a book of advice for housekeepers from any period of American history and it will forthrightly advise you to develop a system, organize your labors, and make better and more efficient use of your time. It would be foolish to argue that these books' promptings describe the reality of time usage in the home. Most women probably balanced such exhortations against the fatigue and boredom of housework and took their rest where they could find it.

But recent histories of housework and the extent of women's labor at home have concluded that such rest was hard to come by.[67] Some have suggested that technological innovations for the home — refrigerators, iron stoves, vacuum cleaners, for example — rarely decreased women's work. Instead, such devices often either concentrated previously shared "work processes" into less work for some but "more work for mother," or raised the level of expectations and thus the work required to meet them.

Take, for example, the vacuum cleaner. After the new machine, the job of cleaning rugs, once limited to beating the carpet outside once or twice a year with help from other family members, became a weekly and even daily task done by women alone. Where a broom might once have sufficed, or husband and children helped to move furniture, roll up the rugs, and carry them outside, now the housewife vacuumed alone, and more often, to keep her rugs up to the new standard of cleanliness.[68] The significant fact is not the amount of work a given woman may have done, but rather that, under the prevailing middle-class ideology of time usage, *work expanded to fill the time available.* Ever higher standards and expectations demanded investing ever more time in pursuing the ideal, so that the vacuum cleaner's boasted savings in time seems suspect at best.

While acknowledging the sheer mass of the work involved in housekeeping, most historians of the subject have seen the home, with some justification, as free from the sort of clock timing that came to rule industrial labor. Indeed, the most articulate and influential spokesperson of "woman's sphere"—Catherine Beecher—worked hard to ground her provisos about time usage in Protestant theology and natural law rather than in clock chimes or market imperatives. She never added clocks to her home designs, insisting instead that the natural hours of the sun should govern time and life. In the ideology of "separate spheres" that Beecher helped to formulate, the home posed as a refuge from the hectoring press of industrial competition and clock time. Housework is even more resistant than factory labor to the timing, regularity, and repetition scientific managers loved.[69]

But the clocks that increasingly governed the industrial work place began moving further into the home as the century progressed. Even Beecher, in 1869, admitted that "a clock is a very important article in the kitchen." "Clocks are so cheap now," one domestic encyclopedia advised in 1873, "that their usefulness and companionableness should secure them a place in every room in the house." "The kitchen must have a clock," another concluded, "for on it a great deal of the comfort, and some of the good temper, of the family will depend."[70] Illustrations for the Kitchen Garden Association, a national organization dedicated to teaching the principles of housework to four- and five-year-olds, showed children laboring busily under the clock in 1883.[71] Near the turn of the century, recipe books began listing precise cooking times for various foods, in minutes, instead of estimating the time or judging doneness by color or texture. As one home economist put it in 1908, not just any clock but "a clock that is an accurate timekeeper" seemed "a necessity in the kitchen."[72]

By the 1890s the impulse to rationalize labor, save time, and conserve effort had coalesced into a formal educational curriculum, home economics. The house of home economics, built on efficiency, modernity, and scientific observation, threw its doors open to Taylorite principles, and by 1910 the first time studies of household tasks had begun. Christine Frederick, and later Lillian Gilbreth, among others, made their livelihood dispensing ostensibly scientific counsel to the overworked and "ignorant" mother. Frederick, Gilbreth, and their colleagues attempted, with mixed success, to make the home a place of industrial efficiency. As one scientific home manager pointed out in 1917, the modern husband's life "is systematized." "He arranges his time so that he arrives at his office or place of work at a stated hour, faithfully performs the allotted task . . . and at a stated time, returns to his home when his day is done." What would happen, the writer asked hypothetically, "if this standard of daily habit were not required of everyone connected with it?" Home life, she concluded, must follow the same regular, clock-disciplined pattern.[73]

If the ideal twentieth-century home reproduced the clocklike regularity of the factory and the commercial world, then was there no difference between public and private time? Certainly the imperative to save time recurs consistently throughout the written sources on housework. But the home sheltered other ideas besides work for work's sake. Beecher's ideal domestic refuge was moral both in its temporal organization and in its *furnishings;* her model housewife labored joyfully to create a wholesome home environment, not just a well-run kitchen. But industrialization made whole realms of mass-produced goods—clothing and textiles, fancy furniture, chromolithographs, phonographs, pianos, kitchen appliances—available to the middle class for the first time. If these products had little effect on the amount of time devoted to housework, they most certainly affected both the ways that time was spent and the experience and meaning of private time itself.

The spate of household consumer goods made the home a place of self-expression for middle-class families, and especially for women. Domestic advice manuals of the late nineteenth century groaned with advice to women on managing the new abundance: what to buy, where to put it, how to arrange it, how to cultivate your own taste and culture while evaluating others. Mrs.

Henry Ward Beecher (Eunice White Bullard), wife of the famous minister and sister-in-law of Catherine Beecher, published her own household guides in 1873 and 1883. Just as her husband, Henry Ward Beecher, adulterated the strength of his forebears' work ethic in his writings and sermons,[74] so Mrs. Beecher joined her sister-in-law Catherine's industry to the special problems of consumption and taste. "Money," she advised women, "can always buy a housekeeper," or the consumer goods to fill a house. "*Love* only can secure a *home-maker* . . . unite the housekeeper and homemaker—let them be one and indivisible," she urged, "and that union will provide a refuge from all outside trouble."[75] Mrs. Beecher urged that industrial commodities for the home be invested with that most "female" of qualities, love. In the abundant material culture of industrial capitalism, women's nurturing role increasingly combined with a newfound calling as tastemaker. The physical surroundings came to reflect the woman's temperament and character—as one wealthy arbiter of taste, Elsie De Wolf, declared years later, "it is the personality of the mistress that the house expresses."[76]

For women, the traditional emphasis on conserving time in the home, familiar from generations of domestic guidebooks, blended into the new ideas of "efficiency" as embodied in home economics curricula. At the same time efficiency, in the consumer marketplace, became a manifestation of the modern "personality," of the trendy modern individual's search for uniqueness. If clocks in the home paired efficiency, on the one hand, with scientific regulation, they also linked "efficiency" to a sense of modernity, novelty, and individualism—to the up-to-date personality.[77]

Paradoxically, in the home the energy unleashed by the doctrine of efficiency forced middle-class women into creating an illusion of leisure. In the ideology of separate spheres, the home stood as a place of rest for men and nurture for children. The clutter of detail in the Victorian house was supposed to embody cultivated taste and deliberation, gentility and ease, moral uplift and peace. A man coming home from a long day would find its aesthetic effect a soothing balm to his wounded nerves. But maintaining this illusion took time; it took energy, effort, and organization. Creating and bolstering the illusion of leisure brought women both more work and a fiercer commitment to the principles of machine efficiency that governed factories and schools. After about 1880 one of the foundations of the appearance of leisure, ironically, was the clock.

In the context of the cult of taste and refinement, clocks in the home became yet another mode of personal expression, an aspect of personality in consumption. "It is no longer possible, even to people of only faintly aesthetic tastes," wrote De Wolf in 1913, "to buy . . . a clock merely that it should tell time." De Wolf, who admitted to keeping a clock in every room, devoted five pages to selecting the proper clocks in her book.[78] Clocks designed for home use reflected the trend toward personal expression through furnishings. Factory time clocks, even at the height of Victorian ornamental excess, remained remarkably frank and straightforward in appearance. Though ornamented, the decorations never detracted from the clock's timekeeping function. A simple rectangular design, severe and imposing, marked most factory or school clocks, reinforcing the seriousness of their role as regulators. There was, to put it mildly, little room for whimsy in the industrial work place. But in the home, whimsy ran amuck — clocks seemed to fasten themselves like barnacles to the most unlikely places.

The marketplace for clocks offered a bewildering variety of odd or even bizarre styles: clocks in painted swells of porcelain sea foam, or perched on the shoulders of bronze Cupids; clocks in the shape of brushes, sleighs, baseballs, apples, or doghouses. In the trade catalogues of the late nineteenth and early twentieth century, cast figures of the Bard of Avon or Voltaire leaned thoughtfully on elaborately scrolled clock/cushions, and clocks placed improbably in the trunks of plaster trees supported little girls on swings. The riot of detail and adornment in so many designs often rendered the clock itself almost invisible—clockfaces in the center of hanging platters struggled pathetically for light amidst a gaudy profusion of scrolls and flourishes. Over nine hundred such clocks, for example, vied for attention in the Ansonia Clock Company's 1904–05 catalogue.[79]

These clocks, like the mail-order watches described above, made time into an expression of self or an ornament to the private realm. If the world outside answered to synchronized regulating clocks, standard public time, and train schedules, then time in the home, private time, posed as its opposite. Standard time, by 1900, was de facto law in most urban places—unauthorized by statute but honored almost unthinkingly in practice. It set a fixed and dependable standard for the conduct of business and industry. In public life and in the factory, for most men and women, clocks and watches epitomized the need to fit oneself into the orderly and regular system of this standardized time. But at home, time-keepers recalled artistic taste, individual expression, "personality," and even playfulness, just like any other home furnishing. And for rural Americans, at least as their taste is revealed by Sears catalogues, timekeepers paid homage to the idealized natural world. The point is not that clock and watch decoration differed from common decorative motifs, but rather that it did not—clocks and watches became ornamental expressions of individuality rather than grim tools of industry. One would no longer think of buying a clock simply to tell time, Elsie De Wolf insisted in 1913, any more than one would "buy chairs merely to sit on." In the home, clocks epitomized a whole range of alternatives to the necessarily restrictive time of public life.[80]

Although watches were traditionally associated with regulation in public life and work, the connection between timekeeping machines and individual personality began infusing watch advertising as well, co-existing with the earlier themes of control and regulation. "A watch," one typical 1916 advertisement insisted, "is a part of you, like your eyes and hands. It moves and throbs with you, goes with you everywhere, lives as you live—as long as you live." The identification of watches with their owners, familiar from the nineteenth century, continued here. But even after death, the advertisement continued, when the heirloom watch passes to your children, "you live again in their lives." In this theme the watch both embodied and transmitted its owner's beliefs, values, and achievements. "A man and his Elgin," the text enthused. "What pride of possession, what comradeship the words express!" As "comrade," the Elgin began acquiring a personality of its own while embodying or carrying on that of its owner.[81]

Watches continued to be offered as empowering devices, though now in more dramatic terms. In one curious Hamilton offering, a huge, vaporous hand reached out to clutch an anxious-looking, well-dressed man. "Too Late!" the caption warned. "There are certain types of men who haven't much respect for time . . . men of little character, idlers and dreamers, rather pride themselves on their contempt for Time. Sooner or later, that ghostly, implacable hand reaches out and turns them back." "When you come to think of it," claimed the text, "there's something awe-inspiring in having in your vest pocket a mechanism so delicate, so fine, so perfect, that it actually knows and measures Time—the power above the stars!"[82] Hamilton's advertisement, however preposterously, implied that good watches helped their owners to actually control Time. It also pointed, more reasonably, to a Hamilton's capacity to distinguish its owner from other personalities—especially idlers and dreamers. Hamilton sold the watch as a way to get some logic, order, and apparent control over temporal chaos while declaring your personal feelings about the seriousness of time.

In this ad, the watch enacts a paradox: it simultaneously makes its owner both more and less an "individual," both more unique and less so. The watch is a possession that singles the owner out as special, and the watch is also a possession that assures that the owner can't be singled out. Standard time established a uniform framework for conducting public life, using timekeeping machines as a model. Earlier watch and clock advertising, following the framework of standard time, had stressed a close identification, a merger, between time-telling machines and their owners. This merger cost the watch owner—like the little man who turned into his watch—a kind of individuality. But it also provided a margin of safety and a sense of belonging for the worker who "needed no watching because he already owned a watch." Watch and clock advertising from the early twentieth century stressed the watch's power to secure the owner's place in the framework of standard public time, and thus his or her virtue. But it also invested this place in the machinery of standard time with the trappings of individuality and distinction.[83]

In the process of identifying and empowering their owners, watches soon began either to confer personality or to acquire specific personalities of their own. "Out from the lee of the dreadnaught darts the destroyer; gaunt, fearless—all a-quiver with pulsing speed," an Elgin ad breathlessly narrated. "On board these eager hounds of the sea, guiding their every move, invariable under storm, stress or shock, you will find . . ." not venerable sea captain nor brave first mate but "an Elgin watch" at the helm, controlling course and speed. Elgin took the regulating function of ordinary watches and dressed it up in the sailor's rugged wool coat. It promised that Elgin watches would make their owners captains of their fate rather than mere cargo.[84]

"Is your watch a gay deceiver?" asked one 1916 advertisement for Hamilton watches. "Are you part of that inaccurate watch market?" In the illustration, a bespectacled solid citizen looks at

Hamilton Watch Company ad, 1916.

his watch. The watch's face bears human features, and it smiles and winks at its owner, as if tempting him to a night of roistering or a moral debauch. "The only market for such watches," the conclusion argued decisively, is with those "lacking in orderliness of mind." The personality of such watches played games with time — in the illustration the minute hand wavers back and forth, making the time uncertain. Hamilton's ad linked inaccurate watches to sensual tempters, men and women of bad character, while a Hamilton's accuracy and sobriety distinguished its morally upright owner from "the thousands of Americans who year after year carry a deceptive, unreliable watch."[85]

The personification of timepieces reached its absurd height in advertisements for the "Big Ben" alarm clock. There was nothing at all special about the clock itself, which was simply a good, reliable alarm. But in a major campaign begun in 1910, Westclox, the manufacturer, distinguished itself from competitors by giving the clock itself personality. "Be at your desk when the bunch flies in — begin your day with a flying start," exhorted a typical example, brisk with the rhetoric of business enthusiasm. "It's sure to get around to the old man's ears — 'Five minutes early, sometimes ten, it raises salaries' says Big Ben." Big Ben, the ad went on, "is a punctual, handsome and long lived sleepmeter. He tells you just how late it's safe for you to sleep. He tells you just when you ought to start down town."[86]

Any alarm clock might tell you that, but Big Ben possessed that magic twentieth-century ingredient, personality. Nineteenth-century watches shored up or reinforced a shaky character, but the Big Ben alarm clock was a character unto itself — it worked equally well regardless of its owner's burden of laziness or sloth. Big Ben even had a son, Baby Ben, "a bright-faced little chap who tells really truly time and will wake you promptly at any time you say." Baby Ben was designed not for children, as you might expect, but for adults — Ben Jr. was a traveling alarm clock, and

Is Your Watch a Gay Deceiver?

Are you one of the thousands of Americans who year after year carry a deceptive, unreliable watch?

If you are, it's safe to say you don't feel any attachment for your watch.

No man ever conceived a sentimental fondness for a watch that wouldn't keep time. Yet they carry them. Why do they?

Jewelers prefer to sell accurate watches. Inaccurate watches are of no use in any trade or profession where correct time is an important factor.

Their only market is with those so lacking in orderliness of mind that they will endure constant inaccuracies, or the misguided who pay too little, not knowing that 99% of the value of a good watch movement is *not* material, but the time and skill of the workmen who made and put that material together.

Are You Part of That Inaccurate Watch Market?

Are you the unfortunate possessor of a watch from whose use you get little satisfaction and from whose ownership you derive no pride?

If so, indulge us for a moment while we tell you about the Hamilton Watch.

The Hamilton Watch is never made with less than 17 jewels.

Railroads will not permit an employee, responsible for the life or safety of passengers, to carry a watch with less than 17 jewels.

On most American railroads every man on a train crew must have his watch inspected periodically and must equip himself with a watch that does not vary over 30 seconds a week. Railroad men naturally choose their watches with care and purpose.

The majority of railroad men carry Hamilton Watches. Their watches keep accurate time.

Now the question is, do you really want a watch that keeps accurate time?

If you do, tell your jeweler so, mention your desire to him casually, sometime.

Observe his interest. Jewelers are but human. They do not like to risk losing a customer by telling him what he ought to pay to get a really accurate watch. They know that lots of people are likely to resent what they call "being talked into paying more."

But when a man comes along who has grasped the fact that if he wants a fine, durable watch he must pay the price of a suit of clothes to get it, the jeweler will go to any length to satisfy him, and will cheerfully give much of his time to seeing that the purchaser has his watch properly adjusted to his personal requirements and that it *keeps accurate time.*

If you begin to talk accuracy to your jeweler he will begin to talk Hamilton Watch to you.

If, before you see your jeweler, you want a broader familiarity with what makes a fine watch,

Write for Hamilton Watch Book "The Timekeeper"

It has condensed into 36 readable pages the story of what makes a watch worthy of carrying. It shows all Hamilton Models for men and women—from the $12.25 Movement Alone ($13.00 in Canada), or a Movement to fit your present watch case, and cased watches at $25.00, $28.00, $40.00, $50.00, $80.00 and so on, up to $150.00 for the Hamilton Masterpiece in 18k extra heavy gold case. The book will be sent free on request.

Hamilton Watch Company
Dept. J   Lancaster, Pennsylvania

When you unroll your watch, do you have to guess where the minute hand belongs? Does your watch mock your intelligence with constant inaccuracies?

Hamilton Watch

"The Watch of Railroad Accuracy"

Hamilton watch ad, 1916. Here the watch is clearly developing a character or personality of its own—bad watches tempt to indiscretion, good watches keep their owners running with "railroad accuracy."

his childish prattle seemed appropriate speech for an infant timekeeper.[87]

In Westclox's ads the product, the timekeeper, literally assumed a life external to the person using it. If the old relationship between the timekeeper and its owner was that of a mirror, Big Ben and his tiny spawn acquired efficient, cheerful, independent personalities of their own. Through sheer force of personality they persuaded rather than compelled, like good, efficient salesmen. Waking up to Big Ben was like being awakened by a friend, a useful friend with a brisk, efficient, modern personality.

These last examples extend the tension between individuals and machines apparent in earlier ads. On the one hand, by making watches and clocks servants of human personality, an expression of self, they empower their owners and grant them a certain freedom. With a Big Ben alarm or Hamilton watch, *you* didn't *have* to be punctual—the timekeepers assumed the burden of punctuality for you. But on the other hand, by granting watches a separate, external identity, making them, literally, sentient beings with character and personality, these ads strengthened the mechanical timekeeper's autonomy and power. In a sense, they give the machine life. In the ads cited earlier, the watch and its owner merged. Their relation of mutual dependence and shared characteristics—regularity, punctuality—encouraged a fusion of identity. But in these last few examples the watch and its owner split away from each other—the watch and clock become separate from their owner and more independent.

The progression described here, from clocks that merge with their owners to clocks that express their owners' individuality, and finally to clocks having personalities of their own, parallels the evolving role of time in daily life. In the 1880s, the date of the earliest ads, accurate, standardized public time was still a highly uncertain quality. Standard time zones had just been introduced, and were strongly resisted in many areas. These early ads suggest an urgent need to integrate oneself into a system of public order whose most obvious symbol was the watch. Other examples, like the illustrations in the Sears catalogue, suggest a pointed hostility to clock time and its implications. These last examples reflect the partial hegemony of standardized public time, and the simultaneous development of a private, personal time unhedged by schedules, timetables, and machines. In sum, these ads suggest that Americans granted the authority for time to mechanical clocks, gaining in return a limited range of genuine psychic freedom. In the next decade, the tension between gaining control over time and surrendering control to time would repeat itself in that most uniquely modern art form, the motion picture.

V

Therbligs and Hieroglyphs

Standard time helped establish new ways of organizing life. Telegraphed time signals, automated signaling systems, synchronized "master and slave" clocks—all these offered new templates for governing work, leisure, and education. These new systems were linked with the problems of self-discipline and social surveillance in an industrial society—in essence, with defining the self in relation to others. While supposedly decreasing anxiety and confusion over time, by making time measurement automatic, standardized time and the innovations it inspired only increased the pressure to fit in with the new framework of clock-based organization. None of these changes went unopposed. Workers resisted Taylor's dismemberment of time; farmers resolutely identified their timepieces with rural life; women turned utilitarian timekeepers into aspects of self-expression. Americans engaged in an ongoing and contentious dialogue about time and its meaning.

From the last decades of the nineteenth century to the first decades of the twentieth, the fascinating history of the American movie offers one example of how this dialogue might be resolved. Motion pictures, following the pattern established by standardized time, offered new models for organizing not just work and life but thought. By World War I, movie makers had learned that by manipulating time on screen they could manipulate the imagination. And the structure of this "movie time" duplicated the themes, concerns, and methods of Progressive politics. The history of movies dramatizes standardized time's profound effects on American culture, politics, and thought.

Peering back into the origins of the motion picture, historians usually turn up the eccentric figure of the English-born photographer/artist Eadweard Muybridge. In the 1870s Muybridge was making photographic records of California for the U.S. Coast Survey, where he met Leland Stanford, the railroad baron and sometime horse fancier. Stanford had just bought a fast thoroughbred, Occident, and he wanted an answer to an old artist's question: When a horse gallops, do all four hooves ever leave the ground at the same time? They seem to, but horses weigh a lot, and they looked silly painted floating in midair. Stanford had supposedly bet a friend that all four feet did leave the ground, and he hired Muybridge to settle the question.

The photographer placed a series of cameras a few feet apart in a row. As Occident galloped down the line, a trip wire triggered each camera in sequence. When he developed the sequence later, Muybridge saw that Stanford had won his bet. In several pictures the weighty horse floated serenely and convincingly in midair, all four hooves off the ground. With Stanford's funds, Muybridge began a series of similar experiments on animals and people. In

1881 he published the results, *The Attitudes of Animals in Motion*, as a tool for artists who struggled with the representation of movement.

Muybridge's novel technique fixed the horse's gallop in fragments of time, images no artist had ever seen before. The unprecedented pictures, each a frozen instant, allowed a careful analysis of the individual components of motion—movements too fast for the eye to follow, like a horse's flashing hooves, could be subject to careful, objective scrutiny. Within a few years he had developed the "zoopraxiscope," a machine for projecting a series of his photographs in rapid sequence. Muybridge's zoopraxiscope, the forerunner of the motion picture projector, reconstituted the dry photographic elements of a horse's gallop in the quicksilver of light. Now a celebrity, Muybridge traveled first to Paris, where he observed the closely related work of physiologist Etienne Jules Marey, and then to Philadelphia, where he began new studies of human figures with the painter Thomas Eakins in 1883.[1]

In 1883, Philadelphian Frederick Taylor was pondering the relation between time and efficiency, trying to find a "standard time" for any job. Autumn of 1883 saw much discussion of time in the city's newspapers. The railroad standard time zones, a new and novel proposition, promised to link individuals and regions under a single, unified clock time. It would eliminate the multiplicity of customs, traditions, and improvisations that governed time. The goal of standardization linked Muybridge and Taylor as well. Taylor began his time study techniques at Philadelphia's Midvale Steel just as Muybridge began his Philadelphia experiments, and it seems very likely Taylor must have known about Muybridge's work. After all, the two men shared certain approaches to the problems of their respective disciplines. Just as Taylor sought a "scientific basis" for quantifying and regulating labor and human movements, so Muybridge hoped, with Eakins, to find a "scientific basis" for objectively analyzing objects and movements in time and space.

Standardized public time offered Taylor, Eakins, and Muybridge an intellectual precedent that suited each's exacting bent. It fixed and objectified time—a traditionally illusory quality—into a reliable, quantifiable, unvarying form. Like time in Taylor's stopwatch studies, motion and light might now be frozen, analyzed, and reassembled in Muybridge's photographs, then subjected to rational analysis in discrete, controlled instants. Taylor and his disciples used watches and clocks as a means of breaking down specific jobs into their component parts, then reassembling them in a more productive order. With Muybridge's photographs, a worker's motions might be taken apart for study; so also an athlete's stride might be analyzed piece by piece. The worker might learn a more "efficient" way of working, the athlete gain a better appreciation of his or her technique, and the artist a finer sense of the qualities of movement.

In terms of their use of time the two fields—art and factory management—pursued similar agendas. Muybridge and Taylor never collaborated in any way. But in the urban America of 1883 they shared a cultural, social, and intellectual milieu that was remaking time. It had taken decades to relocate time from nature to mechanical clocks, and decades more to develop a technology that made clock time practical and useful for governing everyday life on a busy continent. And now Muybridge and Taylor were attacking the standard time of everyday events, dissecting it, finding ways to make it more useful.

Muybridge's work, like Taylor's, represented a new consciousness of time—on one hand of time's diminished stature, its arbitrariness, and on the other of its susceptibility to human control. In this sense, Muybridge's photographs and Taylor's practice of scientific management welled from the same source. Art historians have charted the connection between Muybridge and the evolution of cultural "modernism," especially the Cubists' destruction of space and form. Similarly, the connection between Muybridge's work and the development of motion pictures has also been well established. Historians have paid far less attention to the extraordinary connections between standardized time, scientific management, and the later evolution of motion pictures.[2]

Between 1890 and 1920, American movie makers evolved a new form for presenting ideas. They learned to cut up and recombine time after movie cameras recorded it, to impose narrative structure on the "real" time of ordinary or staged events. Analytical control over "real" time created a new medium of communication with extraordinary power—even created, perhaps, a new way of thinking about communication and politics.

From the beginning, motion pictures and the ideal of efficiency kept close company. The American movie industry began with Thomas Edison's 1891 Kinetoscope, a simple box equipped with viewing slot, lamp, shutter, and a short loop of film moving along rollers. Edison saw his invention as an educational tool and an aid to business, a way to demonstrate machinery or sell products. In a competing device, the "Mutascope," viewers watched a sequential series of cards, arranged somewhat like a modern Rolodex file, flip past the viewing window and give the illusion of motion. The Mutascope's backers also imagined their apparatus as a way of showing machine operations to factory owners and employees. But as aids to efficiency, both devices failed.[3]

More imaginative entrepreneurs instead set up "Kinetoscope parlors," urban storefronts offering long rows of the clattering movie machines to a working-class, largely immigrant audience. Most popular in the mid-1890s, Kinetoscope films typically presented brief, documentary snippets of ordinary events—trains coming and going, street traffic, notable incidents of local news or sporting events. Other loops of film offered short vignettes with a mildly salacious or anti-authoritarian cast—kissing, exotic dancing, skirts blown up by wind. Edison began producing films as soon as their entertainment value became apparent. Most of his early output consisted of excerpts from vaudeville and novelty acts. Largely unburdened by plot, Kinetoscope films depended chiefly on novelty and the appeal of the taboo. Like the modern-day "peep show," which they resembled closely in form if not content, Kinetoscope parlors suffered from a mildly unsavory reputation among the genteel. By the late 1890s, their popularity had begun to decline.[4]

After effective film projectors appeared around 1896, the movies revived. For a time motion pictures appeared between acts in vaudeville houses. But by 1900, movies began running independently, in more or less formal theaters or simply in curtained-off sections of larger rooms. These "nickelodeons" gained movies a larger working-class audience. For a nickel, customers saw a program of very short films run continuously. Over half these early nickelodeon shorts came from Catholic Europe, and few showed any special regard for the "Gospel of Success" and the moral values of the native Protestant middle class—"more often themes ridiculed Victorian values." Early American film entrepreneurs also showed a strain of native, small-town morality—deeply suspicious of new wealth, industrial capitalism, and urban excess. But their productions tended toward documentary footage of current events, satires on familiar public figures, or short, simple stories, often laced with slapstick attacks on authority.[5]

Except in cases of documentary, these films usually mimicked the vaudeville audiences' point of view both visually and thematically. They simply presented a filmed record of what the audience saw live, through a single, fixed camera. Within a few years, however, movie entrepreneurs learned to play a few tricks with the camera and projector, speeding up or slowing down the action or reversing the film, producing novel effects that only made the films more appealing to what remained a largely working-class crowd.[6]

The movies' growing popularity, coupled with their vaguely shady character, soon drew the gimlet eye of the middle class. After about 1905, when the phenomenon grew too popular to ignore, middle-class magazines and newspapers "discovered" the motion picture. They began describing the phenomenon regularly, trying to puzzle out its unique and vivid effects. What was going on in those dark rooms? Intrigued and concerned, *The Nation* went into picture houses for the first time, and came out calling the movies "the first democratic art," reverberating "with the very ideas of the crowd in the streets."[7] What made movies so appealing?

Movies cost very little, and they presented no language barriers to their immigrant audience. Both factors undoubtedly helped draw the crowds. But as early as 1908, observers had pointed to the motion picture's power to compress and expand everyday time. Following the lead of the French magician–film maker Georges Méliès, early directors learned how to speed up, slow down, reverse, double-expose, and otherwise manipulate camera and film to produce apparently magical effects. Projectionists too felt free to crank the films at whatever speed they wished, or reverse direction, when the audience got a little bored. In the movies, marveled *The Independent*, "the building of a skyscraper within a few minutes is a feat easily accomplished." "Time exposure pictures" of a local theater under demolition made it possible "to throw on screen a perfect reproduction of the work. The theater could be demolished within five minutes, and by reversing the films rebuilt within the same period."[8]

Pondering the temporal distortions the movies introduced,

Harper's Weekly called the motion picture houses of 1907 "get-thrills-quick theaters." In the "cops and robbers" genre, claimed an account of "nickel madness," "the speed with which pursuer and pursued run is marvelous . . . a hunted man travels the first hundred yards in less than six seconds," while "a stout officer covers the distance in eight." The earliest films violated the expected, common-sense time span and course of ordinary events with startling effect. Movies altered both the speed and the *direction* of ordinary occurrences—apples fell up, people walked backward, flowers grew from nothing in seconds, shattered rubble leaped up and formed itself into buildings. In the movies, one fan hymned, "all miracles are possible, even that most dramatic of miracles, the reversal of the course of life."[9]

The flexibility of motion picture time paralleled film's spatial plasticity. Even in "documentary" films, which shunned the temporal sensationalism of Méliès-type productions, the ordinary boundaries of time and space shattered. In 1909, the *Survey* glowingly described how film brought disparate places and times into immediate and vivid contact. "On an island 2,000 miles out in the Pacific Ocean" the commentary began, "the exiled lepers of Molokai gather daily before the flickering wonders of a screen that shows them the world of life and freedom. Seated in the luxurious saloon of an ocean liner, a group of travelers study the life-like pictures of the places for which they are bound," while "in Iceland, excited Eskimos applaud the heroism of a cowboy who rescues a captive maiden from the redskins." Movies made far-off places familiar. "I enjoy these shows," one patron confessed, "for they continually introduce me to new places and new people. If I ever go to Berlin or Paris I will know what the places look like."[10]

Movies denied the apparent concreteness and stability, the common-sense fixity, of ordinary time and space in daily life. "We have had many . . . inventions to overcome us in recent years," wrote a critic of the cinema: "the telegraph, the telephone, the wireless, the bicycle, the automobile." Each of these inventions helped overcome the limits of space and time, "but none," he concluded, "is more miraculous in essence or has spread over the world so instantaneously as the moving pictures." Movies disturbed the ordinary, standard time of everyday experience. The phenomenon drove this critic to rhetorical heights of exaggeration; since the movies, he insisted, "events have been taught to record themselves, so that Time seems to merge into Eternity. Yesterday is abolished!"[11]

His frustrated charge reflected the peculiar way temporal certainties appeared to dissolve in the projector's light. The first objection, that "events have been taught to record themselves," referred to both the extraordinary realism of the movies and the mechanical nature of movie making. The novel vividness of movie images gave them the aura of truth. Any movie camera set up, turned on, and left running recorded the events before it with apparently objective, unvarnished realism. Even more, the events were largely self-evident—no one needed to describe them, to say what was happening, when and why. Without such a filter of journalistic interpretation, the events simply "recorded themselves." In Kinetoscope parlors, working-class customers had watched nearly plot-free depictions of ordinary scenes—"spectators were quite content with views of factory employees going to and coming from their work, the arrival and departure of railway trains, the passing of street parades, and similar scenes."[12] Simply "events" presenting themselves, these early films eliminated the interpretive distance normally imposed between events and a mass audience; no cultural or political authority explained the events' meaning. The stay-at-home movie fan quoted above could see what Berlin looked like without anyone telling him what he was supposed to be seeing.

The odd claim that "Time seems to merge into Eternity" referred in part to the novel special effects the movies introduced, like the speeded-up policeman or the building that reassembled itself. Such effects rent the fabric of ordinary time, making illogical or impossible events seem real and vivid. Time, space, ordinary physical reality gleefully disintegrated in the hands of a master illusionist like Méliès, whose films reached a wide American audience. *The Clockmaker's Dream*, a typical Méliès short of 1903, began with an old clockmaker falling asleep in his chair. As he slept, the three ornate clocks in his shop turned into fashionably dressed women. When a giant pendulum appeared, the women used it to form a human clock as their formal dress grew more casual. Finally the shop dissolved into a pastoral tableau and the women's clothes into loose, diaphanous robes. Inspired by their beauty, the clockmaker stood to kiss one of the women, who promptly dissolved back into a clock at his touch. The clockmaker woke to find all as it had been. Méliès played sly games with the fixity and seriousness of clocks, and clock time, in everyday life.[13]

Time also seemed to "merge into Eternity" through the missing narrative voice—the interpreter who could say with certainty what the events were, when they happened, and what they meant. An early slapstick comedy, for example, might depict a little boy kicking a policeman from behind several times. Audiences neither expected nor received any information about why the boy kicked the cop, or about what bad results came of kicking cops. Humor, and interest, came simply from the boy's impudence and the officer's frustrated reactions. While it would be wrong to claim such a film had no ideological content, or even no narrative structure, its primary appeal is a simple delight at seeing authority flaunted. With no plot and little story, these short films largely ignored temporal sequence—the normal links between past, present, and future that the middle class at least depended on to make judgments.[14]

Lacking such a logical sequence in time, middle-class critics charged, movies denied history, denied morality, denied even the possibility of rational thought. The motion picture, *Harper's Weekly* complained, "makes no demand whatever on its audiences, requiring neither punctuality—for it has no beginning—nor patience—for it has no end—nor attention—for it has no sequence."[15] Historic time, meaning the relationship between what has been and what will result from it, seemed to disappear; hence the claim that in the movies, "Yesterday is abolished!"

Film historians once regarded the first decades of the twentieth century as a fairly sterile period. An emphasis on simply grinding out novel footage supposedly reduced the movies to formulaic, uninspiring banalities, with "none of the intellectual excitement and creative ferment which we might expect to find at the birth of an art form." This approach wrongly judges early films only in relation to what followed. More recently, film historians have reevaluated the aesthetic and political content of these first efforts.[16] In later years Dada and Surrealist artists would find in Méliès and the early silent film makers an inspirational model for absurdist attacks on bourgeois "reality." Though they certainly lacked the complexity and highly developed narrative structure of later productions, in the sense of the "timelessness" they fostered, the first generation of silent films marked a profound challenge to middle-class values about time and time usage.

"Timelessness" on the movie screen fostered a sense of "timelessness" in the real world of the theater seats. When motion pictures moved into theaters, they kept their working-class patrons and subject matter drawn from public life in urban industry. The films' brevity and unstructured narrative helped make filmgoing a social experience very much like the convivial world of vaudeville and burlesque—many of the earliest films were again simply records of notable vaudeville acts. With no dialogue to miss, silent movie audiences talked, sang, and laughed along with gusto. "Certain movie houses," commented the *Survey* in 1909, "have become genuine social centers where neighborhood groups may be found any day of the week; where the 'regulars' stroll up and down the aisles between acts and visit friends." While suspending the common-sense time of ordinary events, the working-class movie theater also ignored middle-class norms for behavior in leisure time.[17]

Simple plots contributed to an easygoing, informal attitude about time and time usage. The concept of the feature film, a longer "main attraction" surrounded by lesser entertainments, was yet to come. Early movie audiences dropped in on and left the short films at their convenience, without fear of detention by a long, slow-moving directorial opus. "Stay as Long as You Like," one Massachusetts theater reassured patrons in 1906. In the audience and on screen, ordinary time dissolved. Movie theaters at once disdained the clock-scheduled time of work life, the temporal sequence of narrative, and the inexorable flow of Time itself. In 1912 the Mutual Company, a major film producer, plastered New York with billboards and posters bearing its proud slogan: "Mutual Movies Make Time Fly."[18] Time flew on the screen, where ordinary time melted away, and in the audience, where moviegoers came and went as they pleased.

In terms of time, the closest modern analogue to the early movie houses might be the pornographic movie theater. Pornographic movies rarely depict complex plots, and they run continuously, in near-total indifference to structured time. People come and go as they like. The comparison stigmatizes the early movies unfairly, but at the same time it approaches the way in which the early silent films threatened middle-class values, especially concerning time and its productive use. The early movies offered an alternative to the public, standard time of efficiency and factory production—their temporal plasticity both denied and undermined the version of standard time and clock discipline constructed by industrialization over the preceding thirty years.

It was precisely this difference between time at the movies

and the mechanistic, regular standard time of everyday life that gave movies their special appeal, claimed *Scientific American* in 1915. An article on "Motion Picture Magic" began by describing how "many a railroad man learns to dispense with an alarm clock. If his call is for four thirty in the morning, he will wake with a start at four o'clock." The reference to railroad men and alarm clocks established the mechanical character of the modern "time sense." "Plenty of people," the writer went on, "railroad men in the lead—can tell the hour to within a few minutes any time of day without reference to a watch." These men had internalized clock time, and it now structured their lives automatically.

Yet certain things, the article went on, could disrupt this routinized work clock—unusual events, emergencies, or mind-altering drugs like opium or hasheesh. Such things stimulated and sensitized the awareness of time, which life's ordinary routine dulled. "Opium and hasheesh fiends live days, weeks and months of experience," the article claimed, "in a few hours." But while drugs liberated the "time sense," unfortunately "such drugs affect the body in the exact opposite from the way they trick the brain." Drugs freed the mind but paralyzed the body. The motion picture, on the other hand, "does for us what no other thing can do save a drug . . . it takes normal intervals of time and expands them one, two or a thousand fold, or compresses them by the same ratio." Enchanted by movie time, "we leave the theater with wonder in our hearts and admiration on our lips." The article praised films' ability to liberate the mind from its routinized time sense, to break down the standard time of everyday life and restore "the magic of our childhood," lost, presumably, when we joined the work force.[19]

By drawing the distinction between harmful drugs and health-ful movies, *Scientific American* hoped to point out that movies soothed clock-tired minds and aided productivity. If the movies refreshed tired laborers by suspending ordinary time, perhaps that only made them all the more energetic for the next day's work. True, the comparison of movies to drugs betrays some uneasiness about the cinema's effect. But the assurance that movies left the body unharmed makes the movies seem benign, therapeutic; a "safe" form of leisure.

Extending the movies' tentative relationship to industrial productivity, cinema partisans frequently promoted movies as the natural alternative to the saloon. Saloons gave the workingman some much-needed recreation, but at what terrible cost! Soaked in gin, the drunken laborer lost all sense of time and responsibility. But the motion picture, *Scientific American* claimed, offered the same mental recreation without the physically damaging, family-sundering effects of demon rum and the immoral pub.

But the very "timelessness" of the whole movie experience subverted the middle-class version of time and ordinary reality. Middle-class ideology imposed an apparently logical temporal order on life: work hard, don't waste time, be patient, punctual, obedient, and prompt, and you will succeed in due time. Or, conversely, waste time; be lazy, impulsive, and disrespectful of authority, and you will fail. This may be true, or it may not, but early cinema registered a simple and maddening indifference to this temporal structure. As *Harper's Weekly* suggested earlier, an audience of largely working-class immigrants fascinated by timeless movies, unimpressed by the virtue of punctuality, unhampered by sequence, and indifferent to patience, outraged native middle-class values.

Middle-class critics looked at movie audiences and saw instead of a vibrant neighborhood crowd a somnolent, unthinking mass, hypnotized by the flashing light. "Rarely does such an audience betray animation, scarcely ever awareness . . . its dull eyes unresponsively meeting the shadowy grimaces on the flickering film," worried *Harper's*. Theater critic Walter Eaton charged that "motion picture audiences sit hour after hour without smiling, without weeping, without applauding. They sit in solemn silence in a dim dark room." Like *Scientific American,* these critics also saw movie time as druglike, trance-inducing, but without the positive therapeutic effects. *Harper's* especially viewed the movies of 1913 as a drain on both the minds of its audience and the productive resources of the nation. "You wonder how it can be possible, in an alleged busy world . . . to assemble daily, for long blank periods, so many people who have nothing to do and who are obviously not worrying about it."[20]

Just three years earlier, *Harper's Weekly* had described movie audiences' boisterousness and loud interaction with the screen. What accounted for the difference? It might have been only a reflection of the writer's hostility to the movies — he may have seen only what he wanted to see. But audience behavior may in fact have changed drastically by then, thanks to the rapid rise of the narrative "feature" film, and its connection to time discipline and social efficiency.

———

The noisy debates about moviegoing that filled magazines after 1910 dwelt almost obsessively on productivity and social efficiency — what good, if any, did movies do? In that timeless movie sleep, progressive reformers worried, what dreams did come? Were movies simply a harmless waste of time? Or did their unique temporal and spatial plasticity lend immorality a special allure? What happened when the lights went out and ordinary time was suspended? Did tired workers get some much-needed relief from strain, or simply mire themselves deeper in bovine thoughtlessness? The cinema's growing general audience only increased reformer's concerns. What messages were the movies conveying?

The answer, it seemed, was that the fluidity of time and space in silent films wore away at the temporal foundations of morality.

The early photoplays appeared to lack even the barest skeleton of plot, and their notorious indifference to moralizing conclusions denied "normal" standards of cause and effect—laziness, bad behavior, or gross criminality often went unpunished on screen. Middle-class reformers began working early on to introduce some sense of cause and effect, some temporal rigor, to the silent shorts. Outright censorship offered the most direct form of control over the silent evil.

Movie censors paid special attention to making celluloid crimes connect to their (bad) outcomes. The National Board of Censorship, organized in 1908 to oversee film content, proclaimed that since in real life, as everyone knew, "punishment naturally and fatally follows crime," so it must in the movie house. If the Board had its way, the hypothetical cop-kicker in the imaginary short film described before would now be caught, and hauled tearful and repentant before the local magistrate. By connecting movie crimes to the consequences reformers thought they should have—by showing, literally, that "crime (or laziness, or disrespect) doesn't pay"—movie censors hoped to impose their own version of narrative moral logic on movie time.[21]

Evolutions in cinema style made their task easier. By the second decade of American movie making, the simple novelty of seeing ordinary scenes, or the same scenes at different speeds, had largely worn off. Jaded audiences soon demanded more than simple depictions or manipulations of the ordinary, or rehashes of familiar plots, while reformers began insisting on causative moral sequence. Film makers began finding ways to tell more complex stories. To do this, they had to find new modes for representing narrative.

Edwin S. Porter's 1903 *Life of an American Fireman*, often described as the first American attempt at "story" film, depicted the rescue of a woman and child from a burning building. To modern eyes, Porter's intriguing film handled time and narrative awkwardly. It began with firemen hearing an alarm, assembling quickly, and then racing off and arriving at a burning building. The scene then dissolved to the interior of the building and the mother's room. Mother and daughter pace frantically in thickening smoke, finally fainting onto the bed. A fireman then appears by knocking down the door. After breaking out the window and calling for a ladder, he carries first mother and then child out the window and down, off camera, to safety. Finally the view dissolves to the exterior of the building, and audiences this time saw the whole sequence—the fireman entering the building, calling out the window for the ladder, then carrying first mother and then child down the ladder—again from outside the building.[22]

Looking at Porter's film today is startling; it approaches narrative in ways that hardly make sense to us. Film historians call Porter's narrative technique "temporal overlap." Instead of cross-cutting rapidly between simultaneous scenes, as a modern director would—juxtaposing, for example, the woman in the burning building, the firemen rushing off, smoke pouring from the building's windows, the rescue from inside and out—Porter kept the various scenes of action separate from each other. The key sequences maintained their integrity, their "real" temporal continuity, and the audience viewed them through a single, mostly unmoving camera. Porter in effect had to show the rescue twice, once from each perspective. Film makers had not yet learned how to compose what we take for granted—a focused, omniscient narrative that could combine separate points of view easily. Porter's immensely popular film represented a different way of seeing the world.[23]

Making "modern" films—movies that tell stories in the ways we expect—required new ways of organizing and representing reality *and time*. Porter used "temporal overlap" to show events that occurred simultaneously in different places. He kept them separate, each on its own "local time." A modern director would instead show simultaneity by cutting rapidly between the two scenes of action, ignoring each scene's "local time." In a modern film the "standard time" of the narrative, like the standard time zones that overcame local differences, joins the separate spaces of real experience together. As standard time zones linked distant communities under one time, film makers began learning to knit distinct realms of action together in the fabric of narrative time.

A 1908 film by D. W. Griffith, *The Fatal Hour*, played on this theme. The improbable story concerned a young woman captured, bound, and set before a pistol rigged to a clock. When the clock's hands reach noon, the pistol will fire and kill the woman unless her rescuers arrive in time. Griffith cut back and forth between the woman, the clock, and the rescue party, uniting separate realms of action in one standard time—that shown on the clock-face. Or so it seemed: in fact Griffith accelerated time to increase the feeling of suspense, simply moving the clock's hands forward between shots. An early, limited example of modern editing, the movie used control over time in the film-making process to reinforce the idea of a "standard time" governing and uniting simultaneous events. Control over time, through narrative, valorizes the clock's power.[24]

By 1910, thanks to innovators like Griffith, pressure from middle-class reformers, and the reorganization of film production, many of the narrative techniques we take for granted—cross-cutting, multiple camera angles, close-ups—had come into limited general use. "The motion picture," one admirer realized seven years after Porter's film, "has it in its power to show us what is going on simultaneously in two different places, inside and outside a house for example, or in two adjoining rooms."[25] Film makers quickly learned that compressing or expanding the time of events allowed them to put more complicated plots on celluloid. The new sophistication in handling movie time encouraged the rise, after about 1912, of "feature films"—multiple-reel productions up to several hours long, often with stories lifted from the canon of literature.

Adapting novels and classic stage plays for the silver screen "legitimized" the motion picture; as one magazine put it, after 1912, "college professors didn't have to apologize for going to the 'movies.' "[26] Theater owners turned increasingly to such films, which drew a middle-class audience willing to pay extra for longer, more "serious" productions. The rise of feature films changed both narrative style and the nature of moviegoing. "Features" typically presented far more complicated plots, with dramatic shifts in time, place, and point of view. Missing the beginning of a feature film condemned the tardy viewer to befuddlement. The more complicated stories demanded close attention, discouraging latecomers from leaning over their neighbors for a summary. Using literary plots "uplifted" motion picture content, making middle-class behavioral norms seem more appropriate. As notions of punctuality and decorum adopted from the "legitimate" theater crashed the gates of movie houses, the whole experience became less sociable, more sedate, and more structured in time. If the working-class moviegoer of 1910 came and went at will, chatting with neighbors, his or her middle-class cousin made sure to be on time for the feature film of 1915, and waited silently for the start, clucking at others for their lack of decorum.

As time and space on the screen became more fluid, more flexible, more amenable to directorial control, time in the motion picture theater grew more fixed, more regular, and more defined. In the era of the short one-reelers, motion picture theaters, if they advertised at all, usually listed only their hours of operation— "films shown continuously, noon to 11." By 1915, major "features" were being listed like theatrical productions, with specific starting times.[27] Feature films introduced a paradox central to the experience of the motion pictures: they subjected leisure time to the same standards of punctuality and decorum that ruled work. By putting their audiences on a stricter, less variable schedule, these films imposed punctuality and clock time in the process of freeing the imagination from it. In the hands of a skilled director, cinematic time—the temporal structure of the images on the screen—obliterated the ordinary, inflexible, rigid clock time of everyday experience, substituting the giddy freedom of shifting scene for the clock's unvarying tick. But to experience that freedom, audiences had to show up on time.

Juxtaposing the two experiences of time defused some of the movies' subversive aspect. Showing feature films at scheduled times was like turning a chaotic, lush woodland into a formal park; it replaced apparent disorder with thematic, formal organization, and created a structure or form capable of bearing more complex ideological messages. The experience of movie time, like the experience of nature in the park, became both more intense, more freeing, and at the same time more discrete, safer, further separated from public life and ordinary reality. Feature films concentrated recreation.

And their ability to hold an immigrant audience, to fix their attention on the screen while distilling complicated plots silently, offered reformers special advantages. "The very people who most need instruction," claimed an article on how motion pictures might "Make Good Citizens," "are the ones who patronize the cheap theaters." "This immense audience is more easily reached through the motion picture than by any other medium," argued the Progressive journal *American City*, and the impression movies make "is often deeper than those of editorials or sermons." Reformers quickly realized that moviegoers, especially after the rise of the feature film, were literally a captive audience—simultaneously passive, receptive, and contained. One movie censor reminded his readers of the restless, energetic young people who filled modern cities. Before the movies, he claimed, "thousands of them were on the streets in small groups, searching for friendship, excitement, and mates. Many of the social and moral barriers are let down." But thanks to the cinema, "these young people, nightly, now find their way to the 'movie,' and are at least held under the spell of something more impersonal than each other."[28] Under the spell of movie time, audiences learned the value of punctuality and public decorum. Even the "greenest" immigrant would soon learn to arrive promptly at eight for a promising new "feature," and sit quietly in appreciation.

Feature films separated time more clearly into past, present, and future. They embedded an action in its context, making it easier to trace the chain of events from a given deed to its inevitable results. The multiple lines of action woven into a bad decision could all be portrayed, and each thread followed in time and space as the fabric of crime unraveled. Reformers consistently maintained that only a morally uplifting finale safeguarded public welfare. The National Board of Censorship issued a detailed set of guidelines for movie makers in 1914, again reiterating its particular vision of cause and effect. In all pictures in which "questionable scenes" occur, the guidelines demanded "that the main argument and effect of the picture shall be for good. To this end the censor should see to it that the evil characters in the picture come to harm as a direct result of their evildoing." Regarding drug use, crime, scanty clothing, smoking by women, irreligious speculation, drinking, and drunkenness as social evils, the Board insisted that movies must depict, and depict convincingly, the process of decline that each began.[29] A bad ending to bad business restored the middle-class version of cause and effect in time.

Guidelines like these make cinematic time an instrument of political power—establishing a chain of causation leading from crime to punishment, from drug use to addiction, from scanty clothing to prostitution, makes movie time a disciplinary tool. It imposes a specific temporal structure disguised as real life. Take, for example, the idea that work leads to success. Movies built around this theme would show the process whereby hard work is recognized and rewarded as time passes. An alternative narrative might depict hard work leading to growing exhaustion and finally physical and mental breakdown. In both narratives, a particular idea of work bridges time, spanning the gap between point A—the time when work began—and point B—work's end result. Both tie time's passage to a larger message—on the one hand that work is bad, on the other that work is good. Either alternative subverts someone's idea of the truth, but each one attaches its ideas to and depends on a narrative structure. A more truly subversive film, like the early shorts that working-class audiences watched, almost ignores narrative altogether. Such a film simply depicts work, without an "arrow of time" pointing toward any outcome. By providing movies with the arrows of time necessary to forming political opinions, narrative films restocked the reformer's quiver.

But claiming that narrative served only one point of view would gravely distort the history of American film. The evolution of narrative was neither as smooth nor as uniformly useful for do-gooders as this abbreviated chronology suggests. For one thing, many filmgoers resisted the feature film's dominance into the twenties. As late as 1917, one producer insisted that "the movie fan must be able to drop into a show at any time and get a complete piece. He must not feel that he has to appear at 7 p.m. and stick it out until 9 p.m. to get something. He must be able to drop into two or three shows of an evening and get a complete story at each." Movie theaters were still offering "continuous showings" into the 1920s.[30] And of course even the most articulate, carefully structured film might fall helplessly before its audience's creative interpretation. Simply by dramatizing crime and the exotic, movies made criminals and exoticism glamorous and attractive. But film historians recognize that by 1915 a standard narrative vocabulary had developed, a "classical Hollywood style" that would characterize American film at least until the advent of the talkies.[31]

The rise of feature films suggests both how changing ideas of time derived from industry worked their way into popular entertainment and leisure, and how these new ideas of time served different ideological and class agendas. The short films of the early period, their brevity and lack of plot, made time discipline irrelevant. Narrative movies broke time up and then reassembled it, so it could bear the weight of political persuasion. They embodied complete control over time, yet by putting their audiences on a more fixed schedule, narrative films reintegrated movies into the framework of standard time. In reforming time, narrative films became the entering wedge of Progressive reform—even dreary subjects like hygiene and cleanliness could be dressed up in a romantic plot with general appeal. The story of a young wife, torn from her husband in the prime of life by drinking tainted milk, united the before with what followed it, and theoretically dramatized the need to clean up conditions at the local dairy farm.[32] Film narrative lent even the most mundane political agendas new vigor.

Superficially, movie narrative may seem no different from narrative in novels or plays. Shakespeare's mastery of time, after all, is often cited as one of the hallmarks of his work, and novelists compress and expand time routinely. There are definite similarities. But on a mechanical level, books can never shift from one scene of action to another as fluidly as movies. Imagine a chase scene. No writer could alternate between pursued and pursuer as quickly as a film does; the results on paper would be jumbled and hard to follow.

Even more, cinematic narrative marked a departure in the understanding of time and its relation to reality. A writer cuts up imaginary time in retelling a story that exists only in the imagination. Even if the story a playwright tells actually happened, those events have passed, and the form the author discovers them in bears only a symbolic relationship to the actual duration of the events when they occurred. Two thousand words may express virtually any span of time. But 2,000 frames of film at normal speed equals just over one minute of lived experience—literally, feet per second. To make a narrative, the movie editor cuts up segments of real time on film—mechanically, physically—and then reassembles these segments with a new duration.

With their mechanical control over time, the movies represented a new kind of temporal logic, which by reordering the standard time of everyday life seemed to some even capable of reordering ordinary thought. "These motion pictures are more degrading than the dime novel," insisted *Good Housekeeping*, "because they represent real flesh and blood forms, and impart their lesson directly from the senses. The dime novel cannot lead the boy farther than his limited imagination will allow him to go, but the motion pictures . . . give him first hand experience."[33] Though by our standards these jerky, flickering, silent images hardly seem realistic, early audiences marveled at their realism—when celluloid trains rushed headlong toward the screen, audiences shrank back in fear. *Lippincott's* called the movies "a panorama which is true to life in every detail, because it is reproduced from that wonderful artificial eye, the camera lens. So we know we are seeing actual occurrences—at least, we think we are—and they are so realistic we may laugh or cry or watch the screen with every thought centered on it, as we would read an absorbing novel."[34]

The analogy of film to book acknowledges the growing narrative similarity between the two forms. But at the same time it also points to significant differences. The film is read like a book; it absorbs in the same way, carving up time and space to construct its narrative. But the film exists outside the mind, outside the imagination of the person watching. The viewer is excited, and accepts an illusion of reality, because what he sees in his head when he reads is now projected on the screen—the temporal and spatial freedom of imagination reproduced in two dimensions. Narrative films seemed to make the imagination, with all its boundless command of time, external, vivid, and real.[35] And what's more, movies allowed hundreds of people to share the experience simultaneously.

And this is where feature films represented a new way of visualizing reality—they made the power of imagination, its control over time and space, real at a mass level. Porter's *Life of an American Fireman*, with its technique of "temporal overlap," maintained the integrity of different points of view—the rescue shown first from inside a room and then from outside. The two spaces of action never intermingled in time. Feature films in the "classical style" broke down this integrity and subsumed different points of view in the narrative line. Instead of watching as separate slices of reality presented themselves in turn, audiences were "swept away by an imaginary time flow"—literally by time flowing like the imagination—that combined different realms of action, different actual realities, in one story.[36] The story becomes *their* story, more real than the separate incidents and spaces that make it up, because it incorporates them, breaks down the boundaries in space and time between them. Movies represented a new way of thinking about narrative in time. In this, they seem to have echoed the central agenda of the Progressive era—the desire to create consensus, a unified public opinion and culture, from the diversity of American society.

This last point may seem far-fetched. It would certainly be wrong to say that movies created Progressive politics, or that progressivism invented movie narrative. But it would be equally wrong to ignore the relationship between the two, and the fact that the relationship grew out of new understandings of time. Standard time involved new ways of organizing knowledge, new ways of conceiving the relations between individuals and society. And in the Progressive era, the clearest expression of time's new meaning was the model of efficiency and the machine. Following the relation between movies, social efficiency, and politics through America's entry into World War I will help clarify the way new understandings of time shaped American culture.

In their control of time, movies seemed to offer the potential for a sort of mass psychotherapy. The German-born Harvard psychologist Hugo Münsterberg, impressed by the fluidity of the feature film, emphasized the transcendent quality of cinematic time in his widely excerpted book *The Photoplay: A Psychological Study*. Münsterberg reiterated the difference between cinema time and the standard time of everyday life. He concluded in 1916 that the photoplay worked "by overcoming the forms of the outer world, mainly space, time and causality, and by adjusting events to the form of the inner world"—the world of the imagination. At the movies, he wrote, "the massive outer world has lost its weight; it has been freed from space, time and causality, and it has been clothed in the forms of our own consciousness."[37] If standard time forced an artificial, clock-based time structure on the mind, Münsterberg posed movie time as a more natural alternative, an imitation of the pure freedom of thought and dreams.

But again, paradoxically, the apparent freedom of motion picture time lent itself to a vision of control. Münsterberg came to the movies through his work on scientific management, particularly the relationship of attention and fatigue to efficiency. He advocated scientific career choice based on testing psychological factors like attention span, reaction time, and memory. A fan of the cinema well before 1916, he had recently contributed a series of picture puzzles designed to test "attention, memory, constructive imagination, capacity for making quick estimates, etc." to the Paramount company's screen magazine, *Paramount Pictograph*. His interest in movies paralleled his interest in social and industrial efficiency. In the right hands, he thought, movies might become a therapeutic tool, recharging the imagination by reproducing the workings of the mind on screen, with a commonality that books never approached.

For reformers like Münsterberg, the final difference between movies and books—and this is a crucial point—came in the fact that the movies united the audience in a common imaginative experience, so that all felt the same thing, all "imagined" the events on screen in the same way. At the end of a film, *Lippincott's* concluded, "we may feel so enthusiastic that we unconsciously applaud along with all the others."[38] The audience undergoes a kind of spontaneous union, welded together by film's representation of the form of the imagination just as the film's narrative line welds together distinct, separate realities. Though movies seemed to reproduce the freedom of individual imagination, they also seemed to make everyone imagine things in the same way. Münsterberg envisioned movies as a new art form capable of revolutionizing the aesthetic tastes of millions while promoting psychic and social well-being—that is, of changing the way people think.[39]

To those who feared the movies as a narcotizing waste of time, Münsterberg pointed out that if nothing else the movies invigorated, because "happenings which would fill an hour on the stage can hardly fill more than twenty minutes on the screen," and "this heightens the feelings of vitality in the spectator."[40] An age that made efficiency a secular religion not surprisingly interpreted movies through the quest for efficiency and speed. "Modernism calls for abbreviated action," insisted one magazine, and "in photoplays the plot is unfolded in the least possible time . . . if a stage play requires three hours, in the photoplay it is pictorially told in one hour, and just as effectively." "Have the movies changed us?" asked *Scientific American* in 1917. Indeed they had—corrupted by the movies' brevity and speed, the magazine claimed, the public now "wishes other things to be treated in the same manner."[41] Theatergoers and readers in general, insisted a dramatist, now wanted "Quick Action." "They want to see drama in speed-record time. They want comedy and tragedy that is rapid-fire." The dramatist of the future, she concluded breathlessly, "must tip his pen with radium—and it must leap!"[42]

Carving up and compressing narrative time gave movie makers an aesthetic of streamlined speed. Prospective screenwriters were urged to hone their scripts to the minimum, and focus on compacting the time of events to keep the plot moving. "Any scenes which threaten to be unduly long," suggested a professional script writer, "should be broken up by 'flashing back' to some other scene," or by occasional close-ups which refocused audience attention. *Colliers* reminded hopeful screenwriters to concentrate on action and avoid "wasting words." In a motion picture, proclaimed one movie expert, "everything must be condensed to the irreducible minimum without forfeiting coherency," because for a movie producer, "one factor is all important—time. A producer," he continued, "will spend five minutes in the effort to condense by five seconds the action necessary for a certain situation." *Moving Picture World* urged the would-be scenario writers of 1911 "to cut out every act that does not have a direct bearing on the development of the plot." Only those scenes most productive of an intended effect should remain.[43]

With the rise of feature films, motion pictures returned to their origins in the quest for efficiency, in the parallels between Muybridge and Frederick Taylor. Thirty years earlier Taylor's disciples, armed with stopwatches, had first cut and pasted time according to "scientific" principles. Motion picture scenario writers applied Taylor's now-familiar emphasis on cutting out extraneous, unneeded, "waste" motions to celluloid. Taylor timed a space of labor, separated "productive" from "unproductive" motions, then reassembled the productive motions alone into a standard time for any job. Narrative films mirrored Taylor's understanding of time. They carved up the ordinary, standard time and duration of everyday events and then re-formed it according to the director's particular agenda. Taylor cut out wasted motions and recombined productive ones. In the movies, as in Taylorism, "wasted" footage wound up cut out, while only those most "productive" shots survived. The two techniques, time study by stopwatch and movie editing in frames per second, are directly analogous—each lays out time for dissection and reassembly.

While the director, editor, or censor's cutting and pasting mimicked the time study man's, the organization of production on the movie lot soon came to resemble Taylor's idealized factory. At first, early movie makers had paid little attention to costs, to saving props and materials, to following scripts or dividing labor. Production centered on the director and his team—cameraman, set designer, lighting expert, etc.—all of whom collaborated extensively. Directors and their crews improvised gleefully as inspiration struck them. Divisions of labor blurred in the quest for footage, while without a fixed script directors often wasted time, money, materials, and effort on building sets or shooting scenes only to scissor them out later. But by 1916, when the rise of feature films had raised both the level of movie sophistication and the intensity of competition within the business, major producers like Thomas Ince had streamlined not just the cinematic time of narrative films but the studio time and effort required to make them.

"System and efficiency," insisted the *Saturday Evening Post* in 1916, "have found their way into the manufacture of motion pictures."[44] Major producers, most notably Ince, began working to bring the techniques of scientific management onto studio lots. Early directors proceeded from a simple outline or plot sketch and modified it as they went along. They kept few if any records of specific costs incurred or delays involved in filming any particular scenes. At his Edendale, California, studios Ince substituted a detailed "continuity" script with specific instructions for all employees involved, from director to set designer to carpenter. The scripts described each shot in relation to those preceding and following it, and Ince personally reviewed each script under production, insisting that subordinates adhere to it without alterations.

The innovation was so useful that within a few years Ince retired from directing altogether and settled into the role of "Director-General." Continuity scripts allowed Ince, from this position, to simultaneously control direction (editing and shot choice), oversee and schedule the construction and design of materials in advance, and review future costs. Each script also served as a written record of the film as it moved through the studio on its way to release. In this sense the continuity script echoed Taylor's system of punched time cards, which monitored a given part's movement through the factory. The Director-General's office recalled the routing and planning office Taylor established to oversee production. And like Taylor's machinists, told what to do, how to do it, and how much time to take by an expert, studio employees lost control over the organization and pace of their work to a distant central authority.[45]

While streamlining production and centralizing authority, continuity scripts also helped maintain temporal and causal consistency in movie narrative. As films gained complexity it became harder and harder to unify action from scene to scene, especially without a formal script. Few directors could remember clearly, after a few days' delay, where one shot left off and another began. It would jar even the most unsophisticated audience if, for example, a character entering a room from the right wearing a hat exited the same door from the left, hatless, a few shots later. Continuity scripts let the production manager, and later a continuity specialist, spot such potential inconsistencies in advance, and reorganize camera placement, scenery, and action accordingly. The continuity script made it easier to "tame" and "unify" the spatial and temporal flexibility feature films introduced.[46]

For the audience, the experience of motion pictures as fast, concentrated entertainment related movies to the efficient, productive use of leisure time. Reformers seemed especially sensitive to the relationship between time and efficiency in the movies. The heightened drama that resulted from tight, concise editing lent moral preachments, and the uplifting conclusion, a special force. "While I was studying at Hull House," went a letter to the *Survey,* "I saw in their theater a moving picture of the life of Moses. The dial of time turned back," he asserted, and "it was all there, vivid, true to the Bible story and reverent in tone." Cinema time concentrated and intensified the tale's meaning, so that "even today the motion picture which ran its course in fifteen minutes is more vivid than the story which I have read and taught from childhood."[47] Compressed into fifteen minutes and augmented by the apparent realism of the screen, Moses' story, and the values it embodied, gained new vigor and impact.

Educators were especially anxious to exploit film's potential contributions to efficiency. Münsterberg's comparison of movies to mental processes echoed in the rhetoric of progressive educators, who hoped film might teach lessons faster, more lastingly and more efficiently. In 1913, the *Survey* conducted a symposium on the motion picture in education, inviting prominent educators and reformers to comment on Thomas Edison's elaborate plans for a series of educational films. "Much has been done in the way of scientific management in the field of bodily toil," remarked the editor. "Perhaps Mr. Edison thinks of himself as an efficiency engineer reaching into the mental process—for, of course, time saving and mental efficiency are the very essence of his dreams." In a qualified endorsement of the project, John Dewey of Columbia admitted that certainly in many subjects, "much more information . . . will be given more efficiently, more understandingly, quickly and vividly" by moving pictures. "Two or three performances may easily make an impression more indelible and more intelligent than weeks of reading."[48]

But the most intriguing union of Taylorism, motion pictures, and education came at the hands of efficiency experts Frank and Lillian Gilbreth. Seeking industrial efficiency and an ill-defined "One Best Way" for doing any job, the Gilbreths aimed movie cameras at nearly any human undertaking in hopes of "analyzing an activity into its smallest possible elements, and from the result synthesizing a method of performing the activity that shall be more efficient"—the word "efficient," they added, "being used in its highest sense."[49] By using the motion picture camera to make "micromotion studies" of workers in action, the Gilbreths connected Taylor's principles of time study with the movies' command over time and space. Commenting on a Gilbreth project in Rhode Island, *Current Opinion* suggested that "the foreman of the future will no doubt have to modify his vocabulary, if he is to keep abreast of the times. 'Get a movie on you' will take the place of 'get a move on you.' "[50]

Born in 1868, Frank Gilbreth was working as a bricklayer's apprentice when he first saw the light of motion study. After each skilled bricklayer showed young Frank a different way to lay bricks, he turned the understandable confusion that resulted into the consuming passion of his life—what was the one best way to do this work? In the copy of Taylor's paper on "Shop Management" he came across sometime after 1903, Gilbreth recognized greatness, and when he finally met Taylor in 1907 he venerated the very spot of their meeting to the end of his days.

Gilbreth met and married Lillian Moller, the other great influence on his life, shortly after she finished her M.A. thesis, on Ben Jonson, at Berkeley. His love for efficiency quickly kindled in her, and when Lillian entered the Ph.D. program she switched from literature to psychology at Frank's urging. Her dissertation, eventually completed and published in 1914 as *The Psychology of Management*, attempted to harmonize the principles of psychology with those of scientific management. She drew heavily on the work of several prominent psychiatrists, including Hugo Münsterberg, whose earlier study *American Problems* had suggested a link between the two fields. Together she and Frank produced twelve children and at least as many articles and addresses. Gilbreth was obsessed with efficient production in every sphere. Twelve children, he contended, would pose no strain on a scientifically managed household (especially a household he frequently left on business). The best-selling book and movie *Cheaper by the Dozen*, written by Frank Jr. and his sister Ernestine, depicted their genuinely affectionate family in detail.[51]

Encouraged by his wife, Gilbreth accelerated his motion study work after meeting Taylor. The two collaborated on a *Primer of Scientific Management*, and Gilbreth even helped found the Taylor Society, dedicated to commemorating and spreading its namesake's influence. But by 1910 Gilbreth was growing dissatisfied with obvious inaccuracies in Taylor's stopwatch studies, especially the perplexing problem of "the human element." Stopwatch methods seemed too arbitrary, plagued particularly by errors stemming from the observer's reaction time. Stopwatch study left no visible record other than the observer's description, and besides, it routinely irritated or enraged most of the workers who encountered it.

A habitual moviegoer, Gilbreth broke with Taylor when he replaced Taylor's stopwatch with a movie camera. While viewing some films of his earlier concrete work, Gilbreth realized that movies offered a more objective record of labor and the results of innovation. The pictures themselves could be examined minutely, frame by frame, isolating the components of any set of motions and screening out waste without reliance on memory or snap judgment. Placing a clock within the camera frame, he reasoned further, would give an indisputable record of elapsed time. Even better, films of Gilbreth's innovations gave prospective clients an apparently indisputable record of his method's practicality. Bursting with pride, he described his "micromotion study" approach to Taylor in 1912. But the prima donna of scientific management seemed unimpressed by the implied challenge to his stopwatch.[52]

At Lillian's urging, the disillusioned Gilbreths went into business as scientific management consultants specializing in motion study. To bolster their claims to precision, the couple continually elaborated their movie apparatus. Like Muybridge they placed subjects against a cross-sectioned grid background, to measure the distance their motions traveled. They developed special clocks and watches (emblazoned GILBRETH in bold capitals), designed to show smaller divisions of time — supposedly "up to the millionth of an hour." Finally they attached small blinking lights to their subjects' hands, then photographed them so that the lights left a continuous record of the hand's passage. Called "chronocyclegraphs" by the Gilbreths, these light paths were then reproduced in wire as life-sized, three-dimensional "motion models." Workers in training were to run their hands along these wires and thus learn the proper motion paths. A clock or system of regular beeps kept them moving at the right speed.[53]

The Gilbreth films gave control over time, and dramatized the process of learning. "We not only see ourselves as others see us," they wrote, "but we can go much further and see ourselves *as no one has ever seen us* . . . element by element," instant by instant, frame by frame.[54] In itself this differed little from Muybridge's work, but micromotion films offered a new dimension — they allowed employers "to observe the worker performing the work at practically any speed that we may desire to see him use." Time in a Gilbreth film, like time in the movie theater, could be slowed

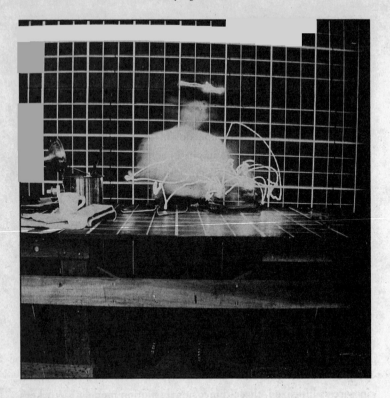

A Gilbreth "chronocyclegraph." The dramatist of the future,
Harper's Weekly had insisted in 1914, "must tip his pen with
radium—and it must leap!" Gilbreth almost literally made that pre-
diction a reality. (National Museum of American
History, Smithsonian Institution.)

down and speeded up at will. The Gilbreths made the comparison
between their studies and commercial films explicit; "those of you
who have made a study of motion picture films," they continued,
"and who have analyzed trick films, where people move far above,
or below, the normal speed of real life, will at once realize the
possibilities in motion analysis that lie here."[55]

The Gilbreths claimed these possibilities were educational—that viewing fast actions at slower speeds made them more comprehensible, while viewing ordinary motions speeded up either made wrong motions ludicrous or planted the ideal of faster work in the viewer's head. In fact, by controlling projection speed, they finalized the comparison of workers with machines that Taylor had begun. The Gilbreths' fundamental goal—the concept they sold to their clients—was stripping workers of their autonomy and subjecting them to machine logic. Watching a Gilbreth subject go through precisely the same motion, over and over again, at any speed the projectionist desired, reduced the subject to a mechanism with a throttle.

The image seems grotesque, and indeed Charlie Chaplin, most notably, would later use some of the same cinematic techniques in *Modern Times* to spoof scientific management with deeply ironic comic effect. And here lies the irony: both examples reflect time's new susceptibility to control. Though we might find Chaplin's use of the technique far more palatable, Chaplin and the Gilbreths both did the same things—cut, pasted, reassembled, retarded, and accelerated fragments of time on celluloid. Their intention differed, but their approach to time was the same.

The Gilbreths edited workers' motion rather than motion picture film; they typically filmed someone at work, revised his or her movements after studying the film, then refilmed that person going through the new, scientifically correct motions. But by breaking down movement their experiments distilled time and motion into discrete units, like the individual shots comprising a film. Just as narrative films reassembled separate pieces of duration, the Gilbreths synthesized the elements of any motion in a new narrative of work. All tasks, the Gilbreths concluded finally, consisted of no more than sixteen fundamental, standard elements. They dubbed each of these sixteen elements a "Therblig"[56] ("Gilbreth" spelled backward, almost).

Frank and Lillian Gilbreth's "Therbligs." The symbols could theoretically be written as hieroglyphic sentences, to describe any job as a sequence of standardized actions. "Avoidable delay" is a sleeping worker, "unavoidable delay" a broken worker, "plan" a man scratching his head.

⬯	SEARCH	(	INSPECT
⬭	FIND	⯅	PRE-POSITION
→	SELECT	⌒	RELEASE LOAD
∩	GRASP	⌣	TRANSPORT EMPTY
⌣	TRANSPORT LOADED	⸗	REST FOR OVER COMING FATIGUE
9	POSITION	⌢	UNAVOIDABLE DELAY
#	ASSEMBLE	⌣₀	AVOIDABLE DELAY
∪	USE	⸮	PLAN
⊬	DISASSEMBLE		

The ridiculous term parodies itself; but the silliness of much of the Gilbreths' work nevertheless fails to obscure their importance. It seems only fair to note that the Gilbreths, even more than Taylor (and like Chaplin), claimed to have the laborer's best interests in mind. The vision of workers helplessly repeating the same motions obscures the real value of some of their projects, particularly the attempts to adapt office machines and procedures for the handicapped. In the context of the era's cult of efficiency, the Gilbreths seemed to be pursuing scientific management's logical course, and though Frank was more than a bit of a faker, their work was taken seriously indeed. Frank and Lillian's central goal—promoting efficiency and the standardization of time, work, and experience—harmonized with many of the basic tenets of the Progressive era.

The Gilbreths viewed "the transference of experience" as their most important goal, and projecting motion pictures of workers in action seemed the most accurate means of reproducing and teaching right and wrong ways to work. Once the "best way" had been established, motion pictures made it possible to reproduce that experience infinitely, to create a standard, universal language of efficiency. By 1918 they approached all tasks as strings of Therbligs, winnowed, organized, and reordered, like the shots in a film, to fit a specific task.

Therbligs broke down the processes of work. If a given job—shoveling, for example—traditionally consisted of eight Therbligs, after the Gilbreths finished studying the task and cutting out unproductive motions, only four might remain. The "first couple of efficiency" hoped to create a standard language of movement for the "transference of experience." They assigned each Therblig both a special color and a symbol. Once time was provided as a line of numbers, ideally any task might be represented on graph paper as a cycle of these symbols. Any motions could be translated into a permanent record of their Therblig components, reconstituted with the aid of movies and wire models, and then reproduced infinitely.

Here again, the Gilbreths nibbled at the edges of one of the central questions of their age. Given the extraordinary vividness of motion pictures, and the control over time and narrative they offered, might not political messages be assembled in the same way, and reproduced infinitely as well? If you could make a universally applicable narrative of the best way to work, could you also construct universally effective moral and political narratives—narratives of the best way to think? As the United States flirted with entry into the European war, the question of the movies' role in political propaganda would become more pressing. Could movies create a standard language of political ideology that could be reproduced exactly, and without limits?

Motion picture studies led the Gilbreths to view all motions as standard, discrete instants in time, each with its own representative symbols. These symbols could be recombined, like the shots in a narrative film, in more productive orders. Similarly, the poet and movie fan Vachel Lindsay speculated on the "hieroglyphic" nature of movies in his 1915 book *The Art of the Moving Picture*. Lindsay compared silent film shots to Egyptian picture writing. Just as, in Egypt, the hieroglyphic of a bowl could stand for a whole range of ideas and associations, so in the movies a single shot conveyed a wealth of meaning. Lindsay inserted hiero-glyphic characters in his text to make the point. "In the Egyptian row is the picture of a throne," he wrote in one example. When you see a throne flashed on the movie screen, he continued, "you know instantly you are dealing with royalty or its implications." Lindsay viewed movies as a potentially unique and profound art form, but complained that too few directors really fulfilled this potential. "It would profit any photoplay man," Lindsay insisted, "to study to think like the Egyptians, the great picture writing people." He advised any aspiring scenario writer to cut a set of hieroglyphic characters from cardboard, and "construct the out-lines of his scenarios by placing these little pictures in rows."[57] The exercise would teach film makers to think in terms of pictures rather than words.

Lindsay's "hieroglyphics" differ from Therbligs in important ways. Lindsay's symbols collapse distinctions, while Therbligs were intended to break apparently unified things apart—one is analytical, the other synthetic. Their similarities come in the ap-proach to time and narrative. Like the Gilbreths, Lindsay envi-sioned motion picture shots as a chain of interchangeable, movable parts. The Gilbreths reorganized work into a movie narrative of efficiency; Lindsay hoped to combine the visual "words" of silent film language—the individual shots—into narrative sentences of greater power, greater ideological "efficiency."

All three tied movies to the same goal: efficient production. As a poet, Lindsay hoped to see more compelling, effective mel-odrama produced on screen, "Whitmanesque scenarios," which could "show the entire American population its face in the mirror screen." Similarly, the Gilbreths claimed that through their movies "the observed and observer become one," and that watching a Gilbreth film created a fascination with one's movements and a compelling desire to make them more efficient. Lindsay, a Populist by sympathy, hoped that movies would "bring the nobler side of the equality ideal to the people who are so crassly equal," and create a higher form of democracy. In his theories the movie crowd becomes, like the Gilbreths' subjects, fascinated with its own movements, producing greater self-awareness and a commitment to self-improvement.[58]

Lindsay and Gilbreth both echoed the common perception that motion pictures represented a new language, a new way of communicating across class and geographic boundaries. "The cinematograph is doing for the drama," claimed *The Independent* in 1910, "what the printing press did for literature, bringing another form of art into the daily life of the people."[59] A nation of immigrants looked with special affection on a medium requiring no words, a "universal language" communicated by visual symbols. Symbols seemed free from the ambiguities attached to words. "If a class of a hundred students learn an oral lesson a hundred mental pictures are formed," went an article in *Scientific American.* "If, however, the lesson is taught by motion pictures the hundred mental impressions are all alike." "Shall we be challenged," wrote one critic in *The Bookman,* "when we assert that [the motion picture] is the language of democracy which reaches all strata of the population and welds them together?" "Can it not be made," he continued, "to bring all degrees of people into a coordinated organism, working in harmony for the greater things of the world?"[60] As a new way of organizing and conveying information, based on new standards of control over time, motion picture imagery promised to unify and standardize perceptions.

Lindsay envisioned "hieroglyphic" movies tied to a mass art that united and improved its audience. "In a photoplay by a master," he wrote, "when the American flag is shown, the thirteen stripes are columns of history and the stars are headlines."[61] For Lindsay, the image literally spoke volumes. At its best, he hoped, the photoplay's universally comprehensible imagery might bridge the barriers of language and class, producing a nation of more passionate, better-educated citizens. Lindsay's widely quoted book suggested that once reconstituted in the movies, the multiplicities and fragments of a nation's life might be reproduced infinitely, with greater impact and unlimited appeal. The entire process depended on control over time, and on narrative structures based on the ability to reorder and manipulate the "real" time of experience.

Lindsay drew his inspiration largely from the work of one man—the master film maker D. W. Griffith, whose wildly popular film *Birth of a Nation* was setting attendance records as Lindsay published. Griffith's epic of 1915 marked a high point in the evolution of cinematic time. Griffith understood more clearly than any of his peers that the fundamental unit of film "was not the scene but the shot." He doubled and even tripled the number of individual shots that had made up the average movie. He used cross-cutting, flashbacks, and close-ups with virtuoso flare, dynamizing movie time with unparalleled facility and vivid conviction. Griffith's innovations completed the transition from short, story-less movies to full-length feature films. His achievement marked the maturity of what film historians have since termed the "classical Hollywood cinema." After Griffith, complex and ambiguous stories could be rendered briefly, vividly, and effectively, thanks to the director's dramatic new control over movie time.[62]

Birth of a Nation also dramatized, as Lindsay understood, the enormous persuasive power of narrative film. *Birth*, based on Thomas Dixon's virulently racist best-seller *The Clansman*, posed the Ku Klux Klan as the heroes of Reconstruction, riding bravely to rescue beleaguered whites from a corrupt and repressive black regime. Knit together by its racist vision, the film fragmented both real and historic time to promote Anglo-Saxon unity. In its famous climax, the sheerest invention on Griffith's part, a galvanized Klan rode en masse to rescue the protagonists, recapturing the South and restoring white unity. Griffith used rapid cross-cutting to build suspense and increase sympathy for the beleaguered whites. As the camera switched back and forth, in close-up and distance, from the two groups of menacing blacks to the rescuing Klan, audiences could not help but *feel* Griffith's belief in the justice of his cause.[63]

The movie encapsulated narrative film's power to create political consensus by manipulating time. It overcame differences in space and point of view—between North and South, between menacing Negroes and noble whites—by subsuming them in a narrative line which unified white society by marginalizing blacks. In Griffith's view, the "birth" of the *real* American nation came only when northern and southern white men reached a consensus on the subject of racial purity. The movie built this consensus by intercutting separate social spaces through narrative. *Birth*'s narrative unity and its vision of political unity both depended on control over time and space.

Although the film evoked widespread protests from civil rights organizations, including the NAACP, audiences flocked to see it in record numbers—*Birth of a Nation* remained the most popular movie in American history till *Gone With the Wind*. Critics lauded its ability to reforge a coherent, gripping political message from pieces of cinematic time. Vachel Lindsay deplored the film's racism, but praised it for fulfilling his vision of the cinema as a unifying force. Inspired by *Birth*'s fantasy of political consensus, he anticipated the jingoist excesses of World War I when he predicted that "in the future mob-movements of anger and joy will go through fanatical and provincial whirlwinds into great national movements of anger and joy."[64]

Birth of a Nation made fantasy seem real. A St. Louis critic claimed that the film was "soaked with great moral ideas" that Griffith and his cast had felt in their hearts. While lesser movies only mimicked emotions beyond their experience, for *Birth*'s cast, "America, the Civil war, the agony of the South and the romantic

idealism of the Ku Klux Klan; all these were real." "Where anything was not real," the critic added in a telling sentence, "the
producers and actors were convinced that it was. The spirit of a
crusade was behind it all." The movie externalized fevered racist
imaginings. In the photoplay of the future, went his overheated
conclusion, "the playing must be done by people who have themselves been gripped by great themes and have an impelling need
to spread by contagion their own overmastering feelings."[65]

The power of Griffith's film came from the juxtaposition of
apparent realism—the documentary quality of film—with the extraordinary emotional and imaginative power that arose from the
manipulation of time. *Birth*'s impassioned urgency overrode the
facts of history as its celluloid Klan overrode "Black rule";
the film's narrative coherence granted it the aura of truth because
it combined different scenes that looked so real. President Woodrow Wilson himself, after a private screening arranged by Griffith,
reportedly termed the film "history written with lightning." Lightning flashed and lit up the dark for a single moment, but Griffith
had frozen a string of lightning instants on film, then reordered
them into a narrative time that flowed toward a vivid political
conclusion. "My only regret," Wilson mused of the film, "is that
it is all so terribly true."[66]

Combining diverse points of view was what seemed to make
movies true for Wilson. In January 1916 he made a short speech
to the Motion Picture Board of Trade which hinted at the similarities between movie narrative and political consensus. Wilson,
who had only recently seen Griffith's film, began by confessing,
"I have sometimes been very much chagrined at seeing myself in
a motion picture." The screen's distortions, its jerkiness, the "apparently automatic nature of my motion," he claimed, missed the
reservoir of thought, compassion, and emotion that lay beneath
the stiff screen facade. But he then described how he had recently
discovered the truth about the "Mexican situation" despite the

fact that each of his many informants lied to him somewhat. "I think," he told the movie men, "the psychological explanation will interest you."

Wilson explained how he pieced together a coherent narrative of events in Mexico by assembling small parts of each account he received—that is, he recombined segments of each separate narrative into a narrative of his own, "the truth" of the situation. Given that Wilson addressed the Board shortly after seeing Griffith's brilliantly edited film, his description suggests a parallel between the President's construction of political reality and the director's cutting and reassembling of scenes and shots. In speculating that the Board might find the "psychological explanation" interesting, Wilson alluded to narrative film's ability to reconstruct a convincing version of the truth from fragments of temporal continuity.[67]

Compare Wilson's speech to D. W. Griffith's explanation of how movies alone could give a true picture of war. Griffith pointed out that no single individual could possibly comprehend the scope of a conflict like World War I—not even the general in command "can see a tithe of it." As he spoke, Griffith was engaged in filming his own war picture, *Hearts of the World*, which like *Birth* used a romantic plot to fuse the fragments of cinematic time into a coherent narrative. Only a movie could unite the disparate aspects of the war into a coherent reality, Griffith felt, for "the camera possesses a thousand eyes and reaches out in every direction, so that it can catch the grand panorama in one instant and in the next it can disclose a minute detail of the most illuminative and atmospheric close-up." The quotation described the construction of movie time, how Griffith took the various viewpoints each shot gave and reassembled them into a "grand panorama," a narrative that he hoped would depict the glory of the crusade for democracy.[68]

Both Wilson and Griffith describe a process of ordering

reality—combining separate points of view into a unified narrative. Like many native-born Protestants of the period, both men worried about American diversity; Griffith wanted America "reborn" in white supremacy, while Wilson feared the presence of immigrants who clung to their native heritage. Irish-Americans, German-Americans, Italian-Americans, Chinese-Americans, "any man who carries a hyphen about with him," Wilson once said, "carries a dagger that he is ready to plunge into the vitals of the republic." As President, Wilson imagined himself embodying an American consensus that united these disparate groups into one will. "I am speaking with the voice of the American people," he insisted; "I express the spirit and purpose of the American people."[69]

As America's entry into World War I began to seem inevitable, the problem of diversity grew more pressing. Wilson knew that most Americans could see little reason to sacrifice their lives and their time in someone else's cause. Why should Irish, Chinese, or German Americans care about England or France? But he also saw America's participation as essential; how, then, to create favorable opinion for war? Wilson, who genuinely hated the idea of war, came to see the war as a crusade for democracy, a mission of moral uplift that would solve the problem of diversity by bringing people—Americans first, then the world—together.

The process of constructing movie narrative is in this sense analogous to the process of building support for the war. Americans would sacrifice their separate opinions in a narrative of growing patriotic enthusiasm—the "story" of the war as a crusade—just as in a narrative film separate realms of experience were edited together by the story. Wilson's approach to political consensus and the construction of movie narrative share a similar organization of time. It would be foolish to claim that Wilson was directly inspired by movies, or that movies somehow either caused the war or caused Wilson to think as he did. But in terms of time and narrative, the congruence between the two forms of communication—political consensus building and movie editing—is startling.

The point is that new ways of understanding and organizing time permeated American culture at every level, from work to politics to entertainment. They offered politicians, film makers, and factory managers alternative models for ordering reality. Over the next five years, as the United States drew closer to war, industrial production and the production of political opinions merged in the cinematic propaganda inspired by World War I.

Commercial movie producers, by 1915 mostly settled in Hollywood, had at first stayed largely neutral on the question of the war, following the mainstream of political opinion as represented by President Wilson. In 1916 Thomas Ince had even produced a pacifist film, *Civilization,* that may have helped Wilson's reelection.[70] By 1916, however, pacifist films like Ince's were rapidly losing ground to "preparedness" pictures which promoted war from behind a thin veil of neutrality. Newsreel footage of the fighting had excited public interest, and propaganda films released by Germany, France, and England had provoked near riots on several occasions. Sensing money, American film producers used the theme of preparedness to present our boys in fictional conflict without blatantly choosing sides.

As products of the industry's maturity, the by now largely standardized narrative techniques of the "classical style," such films encouraged hawkish sentiment, and started anti-German feeling to boiling over. Ironically, anti-Germanism claimed one of its first victims in Hugo Münsterberg. As a prominent spokesman for the German cause, Münsterberg spoke frequently against both the war and American intervention. He found himself increasingly ostracized by his colleagues and vilified in the press as anti-German feeling grew. The strain of public hostility was too much for him: he died of a heart attack in 1916, brought on, according to his daughter, by the pressure of public hostility to Germany.[71] Münsterberg fell victim, in part through the cleverly structured preparedness films, to the phenomenon that fascinated him.

Although Wilson generally condemned anti-German excess and distanced himself from most prewar efforts, he did endorse two quasi-fictional preparedness films, *Uncle Sam Awake* and *The Eagle's Wings*.[72] But when the United States finally entered the war in April 1917, Wilson helped the motion picture industry elbow its way to the front of the parade. Government and the movie industry enthusiastically collaborated in a series of war bond drives, and over the next two years Hollywood ground out a long series of patriotic titles, including *The Kaiser, The Beast of Berlin, The Claws of the Hun,* and *The Prussian Cur.* The latter's climax, lifted from Griffith, featured the Ku Klux Klan riding to crush German spies.[73] The enemy had changed, from rampaging blacks to the bestial Hun. But Griffith's temporal structure—the unifying climax and last-minute rescue—applied equally well to either enemy. Chopping up real time and giving it narrative unity helped unify the audience's feelings about the enemy.

These films proved enormously popular, and certainly helped build sentiment against Germany. From Wilson's point of view, however, they flirted with a dangerous sensationalism. Hating Germany was all well and good, but a progressive war for world freedom had to be tied to uplifting moral values, to a positive vision of change. The narrative speed and trajectory of these films headed not toward moral uplift, but toward unproductive and possibly dangerous hate.[74]

To help build consensus in favor of the war, and direct the manipulation of public opinion more closely, Wilson established "America's first propaganda ministry," the Committee on Public Information (CPI). Under the directorship of muckraking journalist George Creel, the Committee joined censorship of the press and private citizens with an intensive campaign for public opinion.

This "is the war of the American people," Creel insisted, "only they all don't know it yet."[75] To tell them, Creel turned to what he called "the three great agencies of appeal in the fight for public opinion: The Written Word, the Spoken Word, and the Motion Picture."[76]

Creel, a Colorado journalist before his appointment to the CPI, had worked as an extra on some of "Broncho Billy" Anderson's Western films, films which in later years he occasionally screened for his friends and the White House staff. Creel had even written a few short scripts for Anderson at twenty-five dollars each.[77] Besides a love of motion pictures he brought the CPI a familiarity with the power of short, catchy phrases and slogans. Under Creel, the CPI sent forth its famous corps of speechifying "Four-Minute Men" (up three minutes from the Revolutionary War), who condensed war propaganda into punchy, rousing talks exhausting less than 240 seconds of the audience's precious time. CPI guidelines advised them that "there is no time for a single wasted word." The four-minute men worked mostly in movie theaters, usually either preceding a feature film or afflicting the intermission. Cooperation with Creel's fast-talking boosters helped the motion picture industry stave off classification as a "nonessential industry" later in the war, and kept movies in the forefront of CPI operations.[78]

Wilson hoped that the Committee on Public Information could oversee and channel the film industry's patriotic enthusiasm, and at first movie makers cooperated. A specially formed committee of film industry leaders, including D. W. Griffith, pronounced itself ready and willing to work closely with the CPI.[79] But hopes that private industry would bear the burden of patriotic film production faltered in late 1917 over problems in censorship and the distribution of war footage. So, under the leadership of George Creel, Uncle Sam himself went into movie making through the CPI's Division of Films.

At first the CPI merely censored and distributed the copious footage supplied by the Army's Signal Service cameramen. The Signal Service, to document the war effort, had been cranking out as much footage as its limited staff, budget, and equipment allowed, all of which passed to the CPI for review. Unfortunately for Creel, the war resisted cinematography. Heavy, bulky cameras slowed the photographers down, and what few battles the army allowed them to film usually took place at dawn, when low light made decent pictures impossible. Most of the footage consisted of largely static scenes shot behind the lines, especially pictures of industrial and weapons production. But the public craved war news, any war news, and so when he began making feature films Creel attempted to turn this disadvantage to his favor.

The CPI's Division of Films tried to follow Griffith's editorial path, creating unity out of fragments both on celluloid and in American society. "What we had to have," Creel later reminisced with typically cascading ballyhoo, "was no mere surface unity, but a passionate belief . . . that would weld the people of the United States into one white hot mass instinct with fraternity, devotion, courage, and deathless determination."[80] As Griffith carved up and reassembled space and time to promote a coherent vision of war, the Division of Films made a number of multiple-reel "features" by combining Signal Service footage, splicing together shots of vastly different activities to create the impression of a simultaneous, unified commitment to the war in all its phases.

As an official government agency, the CPI felt compelled to restrict itself largely to "real," documentary footage of the war. But from a cinematic point of view the footage was mostly dull. To liven up the drearier scenes and inject narrative momentum, the CPI borrowed some techniques from the commercial cinema. CPI officials enlisted ad man Bruce Barton—"a master," as the Committee put it, "of the short phrase"—to write catchy titles that would help unify and enliven scenes of shoemaking or training camp drill.[81] With Creel's blessing, New York theater magnate Samuel Rothapfel spliced in scenes from other films, used several projectors to simulate filmic dissolves, and reordered CPI short films to present them in the most effective sequence.[82]

Such modifications introduced a greater sense of unity and narrative continuity. With their diverse and wide-ranging scenes of the Allied war effort, CPI publicists claimed, films like *Pershing's Crusaders* and *America's Answer* showed "how every man, woman and child is helping Uncle Sam."[83] In his address at the premier of *America's Answer,* Creel attacked dissenters and pleaded, "let 110,000,000 people stand as one" in determination to win. "America's answer," he insisted, "cannot be made by any single class or by any single endeavor."[84] To spur industrial production for the war, CPI films adopted techniques from commercial movie making. With these editorial strategies, the CPI tried to present an image of unity out of disjointed footage, and increase the very factory production and unity it depicted.

But in comparison to Hollywood efforts, these CPI films suffered gravely from the lack of a coherent narrative structure and unified editorial control over time. Though an energetic campaign of arm twisting guaranteed CPI films a reasonably enthusiastic audience, at least one otherwise favorable review asked, "why not have woven a little heart interest story through the genuine scenes of the front?"[85] Next to their commercial competition, surviving CPI efforts seem disjointed and dull. Throughout the war, Hollywood issued a steady stream of anti-German films using the standard cinematic vocabulary of the "classic style." "These film dramas," the British *Fortnightly Review* recalled later, "had the qualities we have been accustomed to expect in the work of the great American combinations . . . the scenarios were constructed according to the recognized rules, with 'heart interest' and 'punch' and all the rest of it."[86] But "the Signal Service films and CPI productions," as movie historian Terry Ramsaye remembered them, "resembled a history of the war about as much as a scrapbook resembles a historical novel."[87] In other words, they had succeeded in freezing and breaking up time—like a scrapbook—but not in establishing the narrative continuity and momentum necessary to engaging the viewer.

The CPI tried, with mixed success, to exploit the temporal vocabulary of commercial film. Without continuity scripts and centralized directorial authority, the CPI films failed to knit their temporal components into a coherent whole—they lacked "punch." Without plotting, without a consistent narrative thread, they lacked the "heart interest" that skilled directors used to bind the elements of a story together. They remained, as Ramsaye observed, a scrapbook patchwork, never taking on the illusion of reality. To overcome the problem and make CPI productions more like their Hollywood counterparts, toward the end of its brief career the CPI's Division of Films began writing its own scripts through a scenario department, then offering the scripts with approved footage to commercial producers. "The idea," the CPI wrote to James Hoff of *Moving Picture World,* was "adapting to the screen the idea that is back of the Four Minute Men," a brief, condensed, coherent message of patriotism.[88] Before many scripts were completed, however, the armistice finished both the scenario department and the CPI. When the Committee disbanded hastily in 1919, most records of the department's work were destroyed. But in its brief career the CPI had grasped an essential fact: that restructuring time on the screen translated directly into political power.

A more detailed record of time's new role in political persuasion came from the even briefer military career of efficiency expert Frank Gilbreth. When war broke in 1917, Gilbreth offered his services to the government. "If you do not know how to use me," he wrote Washington, "I will tell you." The Army assigned the freshly coined Major Gilbreth to Fort Sill, Oklahoma, and the

production of training films. Gilbreth set to work at once repli-
cating his laboratory environment. He suggested putting cross-
sectioned grid backgrounds in each film, and insisted that "we
have *at least* one clock in every film . . . if there is any unused
background or foreground or space anywhere in the picture."
Gilbreth wanted clocks in every shot to drive home the need for
efficiency and conservation of time. "This is the era of waste
elimination," he wrote to a fellow officer. "Time must be 'spread
out on the table and examined' before it is thrown away."[89] Film
editors spread time, in the form of movie footage, on the editing
table to be examined; Gilbreth tried to use motion pictures to
examine soldiers' chores the same way.

Gilbreth focused in part on speeding up ordinary military
duties by dissecting the time they usually took. He used "Therbligs" to show soldiers how to break down and reassemble their
weapons in less time, or how to aim and fire more productively.
He proposed making micromotion studies to find the most efficient
motions for a bayonet thrust.

But he also concerned himself with questions of duty and
obedience; how to march more efficiently, how to salute faster
and more uniformly. Prospective captions for one of his films
defined discipline as "the instant and willing obedience to orders."
Micromotion dissections of time, a subsequent Gilbreth caption
pointed out, showed that "the average man can come to attention
in 75/100 of a second. Watch this man and see that he comes to
attention in 60/100 of a second. Can you best it?" Fast-saluting
soldiers were more efficient in their obedience.[90]

Dissecting soldiers' motions formed a substantial part of his
work, but Gilbreth also used the special capabilities of movies to
make more obedient, loyal, and willing crusaders for democracy.
He borrowed from psychology in an attempt to make his films
more involving—"we need a psychologist here," he wrote his wife,
"to put 'interest' in these films," and "make the audience partic-

*Still photograph of a soldier under study for one of Frank Gilbreth's
training films.* (National Museum of American History,
Smithsonian Institution.)

ipate." Better training films, he assured his superiors, would require "a humorist, a jingle writer, and a scenario writer."[91] He
adopted cinematic techniques that made his pictures more persuasive and more gripping, including progressions from distant
shots to close-ups that he thought built curiosity. Gilbreth suggested using captions that forced soldiers to respond. "It was his
suggestion," a colleague later wrote, "that we commence each reel
with what he termed 'hate pictures' showing the atrocities of our
opponents." "It was his thought," the officer continued, "to make

each headline short and peppy" and "insert a great many human interest scenes" to "hold the attention of the soldier." "It was his thought to use slow motion pictures" and "insert a great many close-ups in each reel."[92] Gilbreth tried, by restructuring the temporal flow of information on film, to enter the minds of soldiers and bring back the prize of loyal obedience. Like Creel, he sought a more "efficient" means of persuasion, and tried to translate control over movie time into control over thought. But serious illness, and a subsequent discharge, cut short his military career after a few months, and Gilbreth never made most of the films he planned.[93]

Gilbreth's army career encapsulates the government's attempts at movie production—brief and halting, yet potentially of enormous importance. Private industry produced far more effective war films, or at least far more popular ones. The examples of Gilbreth and Creel really prove nothing about the success or failure of government propaganda. Nor are they even the best example of the era's obsession with conserving time—the four-minute men are probably a much better one. What they do suggest, however, is how Americans' new understanding of time offered new modes for representing reality.

Standardized public time established a universal medium for ordering communication in business, a standard located in clocks. Standard zone time, transmittable by electricity to subsidiary clocks, turned synchronized clocks into models for social organization and factory production. Men like Taylor and Muybridge began carving up that standard time, subjecting it to rational scrutiny. Taylor cut out non-essential movements to boost industrial efficiency, and Muybridge turned continuous motions into instants, seconds, frames of time, to increase the artist's knowledge of movement. But with movies, experiments on time entered the realm of the mind, and the ways information about the world was conveyed and received. Creel, Taylor, Lindsay, Münsterberg, Griffith, the Gilbreths; each sought a universal medium for ordering ideology.

Historians have characterized the middle-class political culture of the early twentieth century as a response to diversity, an attempt to rebuild an imagined consensus ravaged by industrial culture and immigration. Movies began as a largely immigrant phenomenon—their silence transcended the barriers of language. "Motion pictures," one critic suggested in a 1918 article on "Motion Pictures and Democracy," "might have saved the situation when the Tower of Babel was built." But now, during the war, he continued, the motion picture "is helping to mobilize the various points of view of the allied countries and marshal them into a single front."[94]

When Edwin Porter experimented with movie narrative, he resisted working time with too heavy a hand. "In the early days," recalled an article on war films, "only the pictorial values of motion pictures were grasped." Later, "its power of narration was discovered and made use of," and "its fact uses became apparent," till "finally argument and persuasion came within its scope." Now, in wartime, films of American unity and commitment to the war "are sent over to disprove German claims and set America in the true light . . . such pictured facts cannot be ignored."[95] Movies took the diversity and heterogeneity of American opinion and appeared to weld it, as Creel insisted, into "a white hot mass" united by the projector's light. "Influencing and binding all men," one movie fan asked at the height of World War I, "who shall say what levels of common thought and achievement shall be attained through the medium of the motion pictures?"[96]

If standard time offered a way to reorganize society on the clock's model, then movies offered a way to reorganize mental processes on a model of efficiency and control over time, a model of continuity and unity streamlined out of fragments. It was, like

Taylor's vision of clockwork factories, a machine model. "If Uncle Sam can teach, write history, plan strategies, argue, persuade and convince by means of motion pictures," went an editorial in *Current Opinion,* then why not create "a cinema university . . . reducing the length of college courses from four years to six months; cinema newspapers to be glanced at in handy assembly halls," and finally, "picture libraries, where they flash facts instead of handing out a book."[97] In the quest for speed, efficiency, and unity, what "facts" would be flashed? The facts themselves are interchangeable, but the underlying structure, the fragmented time recombined into narrative, is still with us today. Do movies mimic our imagination? Or have we come to think in movie time?

VI

The Golf Stick

and the Hoe

"**M**odern life," the pioneering sociologist Thorstein Veblen observed in 1916, "goes by clockwork." Standard railway time and clockwork machinery had replaced "the mechanism of the heavenly bodies," Veblen declared, and now "the discipline of the timepiece" paralleled "the discipline exercised by machine process at large in daily life." Veblen summarized the clock's emergence as the ruling principle of public life, and described the unconscious clock dependency industrialization required.

Industrial nations, Veblen concluded, developed a dual sense of time: one sense of time public and overtly clock-based; the other private and ostensibly "natural," but in fact every bit as bound to the clock. We now use mechanical clockwork as the pattern for governing ourselves, Veblen lamented, and in what he thought a futile attempt to escape "the machine-like process of living," more and more often we resort to "vacations," and the false promise of a "return to nature." Though they masqueraded as a different experience of time, vacations, Veblen felt, only strengthened the clock's control.[1]

When Veblen wrote his gloomy commentary, the word "vacation" had only recently acquired its modern meaning. Noah Webster's 1828 *American Dictionary of the English Language* defined the word primarily as "the act of making void," secondarily as "the intermission of judicial proceedings" or "the regular studies" of a college or school. Webster did include "the intermission of a stated employment" among his definitions, but this hardly corresponded to our modern sense of an institutionalized break from work itself. The word began acquiring its current meaning, "a period of relaxation from one's customary occupation," with the rise of industrial employments in the late nineteenth century. The *Oxford English Dictionary* describes "vacation," in our sense of "a holiday," as "chiefly U.S." and dates its appearance to around 1878—its emergence corresponds, more or less, with the rise of fixed hours of work and standardized public time.[2] The modern idea of the "vacation" makes sense only in a society with a sharply divided understanding of time, a society which pits clock-based public time against a more natural, pastoral alternative.

This divided sense of time characterized the revolution in leisure that accompanied the new century. Preindustrial societies enjoyed less of a distinction between "work" and "rest." Far more than they do today, the two varieties of experience, "labor" and "life," meant nearly the same thing—they intermingled constantly in the course of living. A wise and diligent farmer, finishing one task, went straight to work on another, and even at rest, the farmer remained a farmer; there was relatively little sense of "time off" from *being* an agriculturalist. But industrial labor, along with wage work and clock-time discipline in the factory, brought a heightened sense of distinction between work time and leisure. Actual time off, by our standards, often seems to have been pitiably brief in the late nineteenth century, especially for the poor. Still, industrial labor gave the growing class of wage earners a new kind of "free" time. Time off after the whistle blew at the end of the day could be spent as one wished—to the eternal chagrin of middle-class reformers, appalled at the spectacle of unstructured idleness.

Along with the increase in leisure came new activities for "free" time—movies, phonograph records, travel by train and later by car, a virtual explosion of consumer goods. A few critics, like Veblen, worried that these things all revolved around machines and mechanization. Machines for leisure, or leisure products made by machines—in embracing machine culture, did people themselves become machinelike? But by the 1890s, the fruits of industrialization had grown too tempting for even the most self-disciplined to resist. As the middle class found itself with more free time, more money to spend, and more places and things to spend it on, some of the old taboos against enjoyment in leisure time began collapsing. As they discovered leisure, "respectable" citizens threw off much of their inherited suspicion of pleasurable indulgence. By the late nineteenth century, to a far greater extent than before, conspicuous self-gratification had become the credo of the successful and the envy of the urban middle class.

But these new leisure pursuits mixed pleasure with peril. Old constraints against idleness lingered, and the middle class's discovery of indulgence raised unsettling questions—did the new emphasis on self-gratification herald the collapse of social order and personal discipline? How could society justify uses of time that seemed to produce nothing but empty pleasure? As the middle and upper classes adopted forms of leisure once confined to those beneath them in the social scale, they tended to "moralize" leisure pursuits and shape them into more comfortable forms. Most commonly, leisure time was tied to production. Industrial work enervated people, tired them out and left them pale and listless. Workers of all kinds who enjoyed themselves after work would revive, the argument went, and return to work the next day refreshed. As the history of motion picture narrative shows, techniques borrowed from industry served to make films seem like a socially efficient and productive form of leisure.

But besides the specter of Veblen's mechanized work force, this argument raised other problems. Could mechanized leisure ever truly offer human "re-creation"? And how could one judge a society that lived for transitory pleasures, instead of productive work? Did such a society really use time wisely?[3] These contradictory impulses—the desire for play and the feelings of uneasiness or guilt play engendered—revealed themselves most clearly in national debate over daylight saving.

Daylight saving first became a national issue in 1916. It was implemented briefly during World War I, then rejected by Congress after a fierce debate in 1919. While the debate over daylight saving reflected ambivalent feelings about leisure, it also gave evidence of a deeper unease about industrialization and machine time. Like "vacation" in its modern meaning, daylight saving only made sense in a society ruled by the clock. Its proponents argued that like a vacation, it would restore people to closer contact with nature and preindustrial rhythms. But opponents of daylight saving shared Thorstein Veblen's perception that it only reinforced the authority of clocks and machine time. To farmers, whose work still ran by nature's calendar, daylight saving simply made no sense—if you want more daylight for recreation after work, American farmers retorted scornfully, then why not simply rouse yourself from bed earlier?

The question looks deceptively simple and even a little foolish. But in fact it gets at the heart of changes in the understanding of time and their impact on American society. Why not, for example, simply declare that after April 30 of every year, all the employees in your foundry or office should report to work an hour earlier, at seven instead of eight? Then they could leave an hour earlier, at four instead of five, and get home in time for gardening and sandlot baseball. Seventy years before, American manufacturers had done just that—mills varied their hours with the seasons, to follow the sun. Railroad and steamship lines adopted new schedules each summer, following the seasonal patterns of employment and recreation. This earlier approach certainly seems easier and much more rational, on its surface, than convincing a whole nation to perform the strangely illogical, self-deluding task of simultaneously adjusting its clocks just before bedtime.

But by 1915, insisting that your employees report an hour earlier than the rest of the city subjected everyone involved to undue hardship. Suppose the streetcar lines only started running after seven—how would the employees get to work? Suppose their children's schooling commenced at eight—who would get the little tykes ready and see them off? And suppose your job demanded close attention to the stock and commodities markets—in that case, your competitors savored an extra hour of profitable trading while you nervously shagged fly balls. In a complex, interconnected economy that demanded agreement as to time, simply changing the clocks seemed the easiest, least painful way to simultaneously change the habits of millions.

The two alternatives—changing schedules and hours of operation, or simply changing the clocks—stem from two different approaches to social organization, and two vastly different conceptions of time. The first approach, changing schedules alone, says in effect, "time is nature, and natural imperatives govern our use of time." The second alternative, altering the clocks, insists, "time is what the clock says it is, and we can change the clock to suit ourselves." In the first, "6:00 a.m." is defined by the sun's position, in the second by the clock's hands. If the goal is simply more light after work, then either approach secures it. Certainly we can imagine a society that shifts its customs with the seasons— Americans traveling in Italy, for example, have traditionally been puzzled and frustrated by that country's custom of taking August off, and only modern air conditioning allows us to forget the old necessity of altering work patterns in the hot summer months. We don't make these sorts of seasonal adjustments today, because we have machines that help us ignore nature and run our affairs with mechanical regularity. By 1915, most Americans had already become so clockbound that daylight saving seemed the only way to change their habits.

Standard time, obviously, had already posed the same questions about time and nature, and its rapid acceptance in many quarters had also already answered them—for most people, time was simply what the clock showed, and for them, daylight saving made sense. But this is too simple. For one thing, large segments of the population, as congressional investigators were soon to discover, simply ignored standard time altogether. And again, in many places standard time had marked only a small, indeed imperceptible change from the local time of the sun. Years of custom made the standard clock time seem a "given"—by 1915, even those who bothered to wonder how standard time came about probably didn't know the answer. They looked to the clock for time, simply assuming a connection between the clock and the sun. The real point is that daylight saving posed these questions about time and society anew, and in far more dramatic terms.

If daylight saving epitomized the clock's dominance of everyday affairs, it also reflected a new sense of control over time, a perception that human time might be adjusted to suit not just business needs, but pleasure and convenience. As Chapter One described, an earlier age saw time as a sacred quality, not to be tampered with at mortal whim; indeed, one could no more tamper with time itself than one could alter the sun's passage across the sky. Clocks imitated God's time, and tampering with them trivialized God's will. But industrialization captured time and located it firmly in mechanical timepieces. Nearly one hundred years later, the rapid rise of daylight saving reiterates the fact that for many

Americans, in the cities especially, time had been desacralized, reduced from an object of awe to a servant of the desire for recreation.

Yet at the same time, the widespread opposition daylight saving provoked points to a stubborn refusal to abandon the older relationship between time, nature, and the organization of life. Although today it looks like commonsense, the idea of altering clocks to suit some human whim made daylight saving seem both unnatural and almost monstrous to its many opponents. The fight over daylight saving dramatized some of the essential conflicts of the period—tradition versus modernity, fundamentalism versus science, farming versus industry, the austerity of the Protestant Ethic versus leisure and consumerism—debates which had at their center conflicting beliefs about time's nature and its use. As an Illinois representative put it in 1919, at the height of the daylight saving debate, "this is a contest between the golf stick and the hoe."[4]

His phrase contained the symbols of two very different ways of life, one rooted in clock-governed work and leisure time, the other in religion and local farm traditions. By the end of World War I the country seemed poised for a sharp break with tradition. The 1920 census startled Americans by revealing that for the first time in the nation's history, more people lived in cities than on farms. True, the census bureau counted villages with as few as 2,000 residents as "urban" areas. But there was no mistaking the prevailing trend toward urban life, and toward the abstract, changeable, clock-based time that governed it. More and more, Americans defined themselves against arbitrary social constructs—like standardized clock time, or notions of "efficiency"—instead of the natural world. The intense counterattack on "modernity" that characterized the twenties aimed at rejecting these secular abstractions and returning society to some basis in natural moral law.

In the twenties, the word "modernity" encapsulated a whole range of the changes wrought by industrialization. It included not only new approaches to scientific inquiry and social reform, but also new attitudes about leisure, sexuality, new styles of dress, new kinds of work and entertainment. As the debate over daylight saving suggested, "modernity" and revisions in the idea of time went hand in hand. But the temporal challenge of modernity transcended daylight saving. "Modernity" revised many Americans' most fundamental religious assumptions—about the earth's age, and the origins of men and women; about the great flood—modernity challenged the very legitimacy of the Bible itself. At the 1925 Scopes Trial in Dayton, Tennessee, tradition and modernity fought one of their most celebrated battles, as Americans asked themselves which priorities should govern life—the priorities of the clock, the machine, and an interconnected economy, or the priorities of nature and the Bible. This chapter ends with a reminder, through the Scopes Trial, that attitudes about time are fundamental to our thinking about ourselves, our society, and the terms we use to organize our lives.

When the railroads adopted standard time in 1883, most of the country followed with little complaint. Or so the railroads told it; in fact, as we have seen, resistance to standard time continued well into the twentieth century. A surprising number of communities simply refused to abandon local time, or switched back and forth from one time to another, trying to find a satisfactory standard. In Cleveland, Ohio, for example, many businesses and even whole neighborhoods ignored standard time in favor of the local sun, well into the twentieth century. The *Cleveland Plain Dealer* reported in 1914 that "sun time is used in many Cleveland factories and many homes," especially in "the West Side and the Newburg section."[5] These neighborhoods kept to the sun, apparently with no adverse effects, while the rest of the city used central time intermittently.

Standard time enjoyed as little success in Detroit as in Cleveland. Michigan adopted standard time early on, but Detroit had wavered continually since 1883, and as late as 1900 the issue of central standard versus sun time still rankled. The city had used local sun time till November 20, when the common council tried proclaiming central time the legal standard. A week later they reversed themselves under public pressure, but by 1907 Detroit had returned, temporarily, to central time. Newspaper cartoonists lampooned the city's confusion. "It would not be surprising," a neighboring editor scoffed, "if the old village returns to its first love, horse cars, before spring."[6] Why the conflict over time in the two cities?

Their problems stemmed largely from standard time's effects on daylight. Central time synchronized Detroit and Cleveland with the railroads, with Lake Erie shipping, and with Chicago markets. But it put both cities approximately half an hour behind local sun time. The ninetieth meridian of longitude, which provided the standard of time for the central zone, passed almost directly through Memphis, Tennessee, well to the west. Using the central zone in effect put Cleveland and Detroit on Memphis local time. By central time, Cleveland sunrises and sunsets thus occurred roughly half an hour earlier than they would have by the local time. Since the advantages of central time—synchronization with Chicago and the railroads—only slightly outweighed the disadvantages of early darkness, most Clevelanders kept central time with no particular enthusiasm, or simply ignored it and followed the sun.

Eastern time, on the other hand, adjusted Cleveland's clocks to the seventy-fifth meridian, approximately Philadelphia local time, and put Cleveland on the same standard as the major eastern financial markets. Even better, eastern time slowed the sun. On eastern time, the sun arrived fashionably late in the morning and lingered pleasantly into the evening. Like Cleveland, Detroit stood almost exactly midway between the eastern and central zones, and so again choosing either zone made a perceptible difference in the hours of daylight. On eastern time, the Detroit sun rose much later—at seven o'clock in the morning, eastern time, Detroit's children straggled off to school in darkness, but had more time to play outdoors in the evening. On central time, the sun rose much earlier—Detroit's children awoke in full daylight at seven o'clock central time, but darkness cut their playtime short.

"We are creatures of the clock," reasoned Detroit's More Daylight Club in 1907; "our habits are governed by it to an even larger degree than we suspect." The founders of the More Daylight Club, Dr. George Renaud and C. M. Hayes, saw that no matter what zone Detroit chose, the clock, not the sun, governed life—schoolchildren woke up when the clock said seven, not when sunlight kissed their rosy cheeks. "We get up in the morning, we go to bed at night, we go to work and quit, and we eat, when the clock says to," they asserted. Since Detroiters answered to the clock and not the sun, casting Detroit into the "fast time" of the eastern zone would give later sunsets, and more light for recreation and play, to workers on regular hours. Why not exploit this allegiance to the clock, instead of suffering thoughtlessly under its heel? With partisan enthusiasm, the More Daylight Club began trying to move Detroit into the eastern time zone.[7]

Detroit's More Daylight Club promoted an early version of that familiar illusion, daylight saving. We all know how daylight saving works today—Spring ahead, Fall back, one hour twice a year—and the advantages and disadvantages it brings. But few people realize that daylight saving has a history, and that its history is inextricably bound up with evolving ideas about time. The idea dates back at least as far as Benjamin Franklin, who twice proposed tongue-in-cheek antecedents of daylight saving for London and Paris.[8] Franklin's sarcastic proposal evoked amusement, not action. But by the time the More Daylight Club came across it, the idea of shifting the hours of work had begun to take hold.

Detroit's early risers drew on the singular obsession of William Willett, a prosperous English builder. Like many fans of daylight saving, Willett loved golf, and regretted the early sunsets that robbed Englishmen of healthful exercise outdoors. As a businessman, he resented wasting fuel to light the darkness. In 1907 he wrote a pamphlet, "The Waste of Daylight," which proposed setting all British clocks ahead two hours each summer. His plan provoked ridicule at first, but gradually gained supporters after Willett sent his pamphlet to members of Parliament, businessmen, "Physical Culture Organizations," and foreign governments, including the U.S. Congress. In Detroit and Cleveland, Willett's idea strengthened the movement toward eastern time. The More Daylight Club's constant agitation brought Detroit to a vote on the issue by 1911, and enthusiasm for the plan spread southward. Cleveland adopted eastern time on May 1, 1914, and a year later Detroit, following a similar protracted battle, set its clocks to "fast time."[9]

In both cities, newspapers endorsed the measure's contributions to leisure time. The Cleveland Athletic Club celebrated the time change with a special party for five to six hundred guests. Two professional dancers from New York "demonstrated the hesitation waltz, Argentine tango and Maxixe with variations and introduced a new one to Cleveland called the Parisian one-step," while two local couples "gave an appreciated version of the cakewalk brought up to date." In both Cleveland and Detroit, commercial firms rushed to exploit the new time: Cleveland papers included a full-page ad headed: "Eastern Time Adopted by these Progressive Concerns." It offered the timely sales pitches of thirty-three different firms, all grouped, however loosely, around the time change. "More time for pleasure," one local store blared, "with the Connersville system of vacuum cleaning." "No matter what time you adopt," suggested a newly opened amusement center, "spend a part of it this summer at LUNA PARK."[10] The advertisements and the Athletic Club's celebration both related the change in daylight to the pursuit of pleasure in leisure time.

But newspapers also hastened to connect daylight saving to productive work and structured leisure. The front page of the *Cleveland Plain Dealer* showed a harried man rushing along behind a lawn mower, over the caption: "One More Hour of Daylight." The picture recalled family life and the responsibilities of an orderly, well-kept home, not nightlife and dancing. One local factory owner proposed dramatizing the benefits of the extra hour of light by staging an exhibition baseball game between the city's amateur and professional players. The Chamber of Commerce lauded the move's contribution to outdoor recreation, and listed "more efficient work in industrial plants" prominently among its supposed benefits.[11]

In 1916, England, France, Holland, and Germany all adopted the "fast time" of daylight saving as a way to promote efficiency and conserve fuel for the war. Impressed by Europe's example, and the measure's success in Cleveland and Detroit, American partisans of more daylight began agitating Congress for a national daylight saving law. Such a law, they argued, would greatly aid productivity, by saving fuel and by giving tired industrial workers an extra hour of health-giving recreation outdoors. Even better, daylight saving would reverse the ongoing trend toward keeping late hours, and return society to harmony with the sun. American politicians began studying the value of turning the nation's clocks ahead.

The Senate Committee on Interstate Commerce met in 1916 to consider "a thorough investigation of all questions involved in the standardization of time in the United States and its territorial possessions." The Committee recognized the fact that without a national legal standard, cities, towns, counties, states, or regions might all adopt any time they chose—as in fact they had done and were still doing—and that such a situation hardly furthered efficient commerce. And the wartime adoption of daylight saving by the European belligerents made the measure's supposed contributions to efficiency particularly intriguing. Certainly the United States deferred to no nation in its commitment to maximum efficiency. The Committee decided to proceed with an investigation into daylight saving.[12]

While Congress pondered, the daylight saving issue dawned in the popular press. There, however, the quest for more daylight met with suspicion. Almost universally, early comment on daylight saving depicted the measure as an ill-disguised attempt to make early risers out of late sleepers. Even its opponents had to admit, as one engineering magazine put it, "that the early morning hours are hours of exceptional efficiency." Germany had adopted daylight saving, after all, and Germany "sets the pace in many lines of efficiency." But most articles depicted the measure as the pet project of moralizing insomniacs. "Stated plainly," declared the *Literary Digest* of daylight saving, "this amounts to a plan for making people get up earlier by telling them it is later than it really is." Most writers resented daylight saving as some sort of fakery, a sugar coating on the bitter pill of morning. The *Saturday Evening Post* called it "a harmless piece of buncombe," but asked sarcastically "why not 'save summer' by having June begin at the end of February?"[13]

The measure's association with industrious living and efficiency made it look like another in an ancient series of nagging imprecations against lying late in bed, a tyrannical imposition by joyless reformers. "The Kaiser has declared that noon shall be called one o'clock," went an article on "fiat time" in *The Independent;* but "we question whether setting the alarm clock of a nation one hour too fast will make people get up earlier in the long run. If they really wanted to get up earlier they could do it now, most of them, without any legislation." "The early morning hours," they suggested, "are highly spoken of by poets who nevertheless are apt, like the rest of us, to spend them in bed." "What special good is going to be accomplished?" asked the *Engineering Record*. "The hours of labor will be neither lengthened nor shortened," and "the actual savings in artificial light," despite claims to the contrary, "is rather illusory."[14]

Daylight saving at first met largely with skepticism and suspicion. The *New York Times* resented the supposition that clocks ruled life, and that tyrannical timekeepers alone could snatch the reluctant sleeper from Morpheus' grasp. "Are we slaves of the clock?" asked an editorial. "Would you get up if your alarm clock told you to, even if there were no sunshine through the bathroom window? Would you go to bed an hour earlier just because your watch said it was the proper thing to do? Does the world move by the clock or by the sun?"[15] The *Times* asked the central question—how far away from nature had society moved? Was American society based on nature or the machine?

One of the more peculiar aspects of the debate on daylight saving centered on the apparent (and apparently welcome) decline of early rising. Changes in American technology, after all, had helped alter not just the idea of time but the schedule of life and leisure itself—electric lights, for one example, brought light and laughter to nighttime hours formerly too dark for all but sleeping. "The fact of the matter," declared *The Independent*, "is that people prefer darkness to daylight for their hours of recreation, and whenever they get money and leisure enough they shift their working day to later hours."[16] So much modern leisure seemed to take place under artificial light—would Broadway lose its thrill by daylight?

Industrial societies stayed up later to consume the fruits of their achievements—"in all civilized countries," insisted *Scientific American,* "there is a tendency to keep later instead of earlier hours, and the daylight saving plan appears to be a somewhat violent plan to combat the instinct, whatever it is, that underlies this tendency." "The real question," continued a second editorial, "is whether it is desirable for the bulk of the population to keep earlier hours."[17] Industrialization may have imposed rigid schedules on work, but it also offered free time after hours, and a dazzling spectrum of new leisure time diversions. Questioning the worth of nighttime pleasure seemed to challenge the redeeming virtues of industrial progress.

By this light, daylight saving looked like a reactionary attempt to beat back change. "It has yet to be proved," declared an editorial in the *Times,* "that a return to the hours kept by hens and savages confers any ponderable benefits."[18] These early critics interpreted daylight saving as an attack on modernity. They resented it as an attempt to regulate life by some health reformer's clock, a thin excuse for wresting more "efficiency" from the vast majority who chose to sleep late. The idea appeared not as a way to use more daylight in the evening, but as a poor ruse to get people out of bed earlier, made all the more offensive by its implicit assumption that people were simply foolish dupes of their bedside alarms. Over a century before, Benjamin Franklin had issued his famous and tiresome maxim "Early to bed and early to rise, makes a man healthy, wealthy and wise." Much of the early reaction to daylight saving suggested that Franklin's descendents still resented the advice.

But by early 1916 a powerful coalition of businessmen and politicians had begun to embrace the idea of changing the hours. The *Times* reported support for advancing the clock among "influential members of the Stock Exchange" who hoped to restore customary relations with the London financial market altered when England passed its new "Summer Time" law. Our present timekeeping customs, Commissioner of Immigration Frederic C. Howe pointed out to New Yorkers in another example, "are a hold-over from when we were more or less an agricultural country." The really modern thing to do was change the clocks — Howe cited the example of Cleveland and Detroit, and suggested that Gothamites could use the extra daylight to "play tennis, go to the country, or to Coney Island."[19]

Manhattan's borough president, the alliterative Marcus M. Marks, formed a committee, including representatives of government, business, and labor, to study the question for New York. By 1917 he had organized the National Daylight Saving Association, a lobbying group including delegates from state and local governments, chambers of commerce and trade organizations, prominent businessmen, and labor organizations. After six months of extensive lobbying in New York and Washington, the National Daylight Saving Association held its first meeting in New York City on January 30, 1917. Over one thousand eager National Daylight Saving Conventioneers converged on the city for two days, "wearing lapel buttons bearing a picture of Uncle Sam turning the clock forward one hour." Speakers on the virtues of more daylight included John Tener, president of the National Baseball League, who spoke on "why fans favor the movement," and the president of the National Lawn Tennis Association.[20]

More significantly, by January 1917 the United States was being drawn ever closer to the European conflict, and in April of that year the country declared war. An obsession before the war, efficiency acquired near-religious status once the war began. Representative Borland of Missouri introduced a bill to begin national daylight saving shortly after the declaration of war. The bill died quickly, but it helped rally nationwide support for the idea. To promote the measure's practicality, the National Daylight Saving Convention printed up and issued mailing cards for concerned voters. Each three-by-five card featured a picture of Uncle Sam with a rifle over his shoulder and a garden hoe in his hand, turning back the hands of a large pocketwatch. The pocketwatch had legs and a face, and wore a doughboy's hat. It exclaimed: "Turn the clock ahead ONE HOUR Uncle. No longer working day for anybody but all in daylight—and you GAIN ONE HOUR for National Efficiency." On the reverse, the cards read: "If I have more Daylight I can work longer for my country. We need every hour of light."[21] In hitching daylight saving to the war effort, its partisans had found their most effective strategy.

While the National Daylight Saving Convention bombarded congressmen with these cards, the Boston Chamber of Commerce, led by department store magnate A. Lincoln Filene, printed up and circulated its own pamphlet on the advantages of daylight saving, "An Hour of Light for an Hour of Night." The pamphlet, which emphasized the need for increased efficiency in war, went to Congress and to commercial organizations in other cities.[22] As a leading member of the U.S. Chamber of Commerce, Filene led that organization's endorsement of daylight saving as well, and won its position substantial publicity. In May 1917, Filene, Marks, and a varied assortment of daylight saving partisans traveled to Washington at the request of the Senate Subcommittee on Interstate Commerce.

The Subcommittee had met to consider S. 1854, "A Bill to Save Daylight and Provide Standard Time for the United States," introduced by Senator William Calder of New York the month before. Marks, opening testimony in favor of the measure, began by emphasizing its potential contributions to wartime production. "The purpose that brings us here at this time," Marks solemnly intoned, "is President Wilson's message that the food supply needs increase." A month earlier Wilson had listed increased food production among the Allies' most pressing problems. Under daylight saving, Marks claimed, "the stimulation of gardening is tremen-

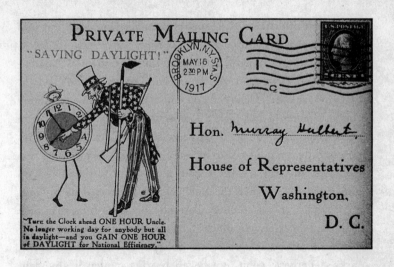

Mailing card from the campaign for daylight saving during World War I. (National Archives.)

dous"—patriotic Americans, spades in hand, would fall out into their backyards and begin digging at once.

Marks even carried a brief letter from President Wilson endorsing "any movement which has the objects of the daylight saving movement."[23] The Subcommittee did hear some testimony in opposition to the bill, from a representative of the railroads who expected the time change to decrease safety on certain lines. But impressed by the potential to increase efficiency, its members recommended that the bill be passed, if amended to take effect on January 1, 1918. When the Subcommittee reported its findings on May 25, 1917, it stressed the innovation's usefulness in time of war, and the relatively minor inconveniences that would result.[24]

Fertilized by congressional and presidential endorsement, Marks's energetic campaign began bearing motley if abundant fruit among the general public. "We have been slaves of the clock," Marks exclaimed; "it is time now to make the clock yield to our needs." In New York, Mrs. Vincent Astor pronounced her approval of the measure, followed shortly by the National Civil Service Reform League and a Dr. Eugene Lyman Fisk of the "Life Extension Institute, 25 W. 45th St." *Automobile,* the magazine of motoring, published a series of favorable comments on a measure that offered more sunlight for motor jaunts in the country after work. The magazine's enthusiasm was tempered, however, by reports that in England, daylight saving had actually *increased* fuel consumption by over-eager evening motorists.[25]

On the more serious side, the move for more daylight gained the support of Samuel Gompers and the AFL. "We urge the inauguration of a daylight saving project," Gompers wrote, "for the conservation of time and opportunity for greater leisure and open air exercise." And the Pennsylvania Senate reflected the anxieties of war when it resolved that since "enemy governments have passed laws advancing the time one hour during the long summer days for the purpose of conserving their resources," so too should Pennsylvania and the nation.[26]

Over the next nine months, the National Daylight Saving Convention issued a series of posters emphasizing the measure's contribution to food production and wartime efficiency generally. Each depicted both Uncle Sam and the doughboy watch that Marks's organization had used in its earlier congressional mailing campaign. "Saving Daylight!" blared one, a poster of Uncle Sam and the doughboy timepiece marching off with hoes and rifles over their shoulders. "Mobilize an extra hour of daylight and help win the war!" In another, Uncle Sam stretched sleepily in his bed while the tireless doughboy clock pointed out his window, where "Germany" and "Austria" forged swords and tilled the soil of "Middle Europe." "Uncle Sam," the caption read, "your enemies have been up and are at work on the extra hour of daylight—when will *you* wake up?" The posters all stressed the importance of staying even with the enemy, and sacrificing at home to win the war.[27]

The idea of the soldier as a watch might seem vaguely discomfiting—it makes the soldier into a machine who tells Uncle Sam what to do. The timekeeper offers a model for patriotic, diligent, early-rising service in the first truly mechanized war—Uncle Sam literally follows the watch. The war required unanimity; as Chapter Five discussed, Wilson oversaw an unprecedented barrage of propaganda and publicity for the war effort, aimed at repressing dissent and uniting Americans in enthusiasm for the fight. Part of daylight saving's appeal lay in its mechanical universality—it affected everyone, regardless of race, ethnicity, or gender, and united them temporally in a measure of sacrifice for the war. But paradoxically, it also seemed to free people from work-place mechanization, and restore a measure of contact with natural time. By the summer of 1917, daylight saving lobbyists had changed opinion on daylight saving in popular magazines from largely skeptical to almost universally enthusiastic.

Encouraged by growing support for the measure's popularity and apparent contributions to efficiency, in January 1918 the House of Representatives entered into debate on Senate 1854, Calder's bill "to save daylight and provide a standard time."[28] Over a dozen speakers in favor of the bill reiterated the by now familiar points: the measure's success in Europe, its boons to efficiency, health, and recreation, and especially to home gardening. A broad coalition of manufacturers, businessmen, and recreational associations supported the idea, its proponents reminded. Form letters, sponsored perhaps by Marks and his organization, demanded the passage of the Calder bill. "What is holding up the daylight saving bill?" asked one. "One million tons of coal might have been saved if Congress had passed this bill a year ago . . . and yet it is allowed to remain dormant."[29] Herbert Hoover, head of the recently formed United States Food Administration, pronounced himself "only too pleased" to endorse a measure so conducive to home gardening.[30] Such testimonials made a persuasive case for saving daylight in time of war.

A few congressmen found these arguments unconvincing at best, and ludicrous at worst. Representative Otis Wingo of Arkansas especially distinguished himself with sarcastic humor in the folksy southern tradition. Apparently serious, Wingo cited a recent article proposing "that Congress provide for a winter thermometer and fix the freezing point at 45° Fahrenheit." Placed in American homes, Wingo continued, the thermometers would read thirteen degrees higher than the actual temperature, and so "they could look at the thermometers and be fooled, and in that way save fuel next winter." The idea seems ridiculous, but it operates on precisely the same principle as daylight saving—the assumption that people obey their measuring instruments unthinkingly. Most of the men who advocated daylight saving, Wingo scoffed, were trying to dupe themselves and their constituents out of lazy habits; they "have not seen the sun rise in twenty years, and they will not see it if you pass this bill." We have more important things to do in this time of war, he insisted disgustedly; but "while our boys are fighting in the trenches we are here like a lot of schoolboys 'tinkering' with the clocks."[31]

Kentucky's Robert Thomas condemned those "modern Joshuas" who sought to alter the planet's course "by any legislative legerdemain or so-called daylight saving device." "This bill," he declared, "was sired by an uplift magazine and will be damned by a majority of the people."[32] The barb struck home — no single magazine had "sired" daylight saving, but the measure was part and parcel of the era's determined emphasis on "efficiency and uplift" through rational legislation. Both men felt uneasy about a measure that depended on rote obedience to mechanical timekeepers, a measure so clearly and unequivocally designed with the industrial wage earner, not the farmer, in mind. But their objections paled before the barrage of testimonials and endorsements loosed by the proponents of daylight saving.

After a lengthy debate, the bill passed the House by a vote of 253 to 40. The Senate added its approval the next day, officially declaring that clocks be moved ahead one hour at 2:00 a.m. on May 31, 1918. The change elated its supporters in New York. Members of the National Daylight Saving Committee gathered at the Aldine Club to watch the clocks change, but otherwise the move excited little controversy and came and went smoothly. "It was as a war measure that Joshua commanded the sun to stand still," recalled the *Times*. If there had been clocks in those days, continued the article, "Joshua might have put the clocks forward an hour or two in order to save that much daylight for the battle." On April 1, the newspaper reported, uncounted thousands of New Yorkers used the extra daylight for baseball, golf, tennis, and thrill-seeking at Coney Island. Little over a month later the newspaper praised the innovation's contributions to social well-being, and asserted confidently that "the idea of abandoning it, ever, in peace or war, never enters any mind."[33]

National magazines registered the same confident approval, with glowing encomiums on the virtues of evening sun. Here, it seemed, was a progressive measure that actually worked. It increased national efficiency, claimed *Current Opinion,* and in addition let working women walk home by daylight; it gave parents a new hour outside with their children, and cut electric and gas bills in half. As the bill's term neared expiration in late October, the *Survey* praised its apparently universal virtues. "Not only recreation in the recognized sense," wrote a Chicago correspondent, "but a general feeling of freedom of life, of neighborliness, of time to visit with the home folks . . . and invite one's soul through the convenient exercise of leisure have been promoted by daylight saving."[34]

As this letter shows, though the measure promoted efficiency it also provided genuine relief from the nagging press of ordinary time, and restored tired citizens to natural pleasures. Daylight saving has made us, the *Literary Digest* concluded, "a healthier, wealthier, wiser land." The measure was not evidence of a mechanized, machine-enslaved society, but rather evidence of control over machines. We have not merely deceived ourselves, the magazine added, but instead as the *Boston Transcript* put it, have "simply demonstrated our moral superiority to our mechanical arrangements."[35] The *Transcript*'s observation touched the heart of the matter—daylight saving took the "mechanical arrangements" that governed ordinary affairs and turned them to more pleasant uses. Supporters of daylight saving used "efficiency" to put their bill through Congress, and depicted a civilian army of fervent war gardeners, patriotically weeding the tomatoes after work. But in descriptions of the measure's benefits the themes of "efficiency" and home gardening largely fell away, replaced instead by glowing accounts of the soothing, relaxing effects of late evening sunshine.

In the grim, urgent, workday world, the enemy time ever threatened to vanish; harried laborers in all fields of work seemed to chase endlessly after ever smaller fragments of time. But not only did daylight saving extend sunlight; it also seemed to unite a whole society in renewed appreciation of nature, leisure, and healthful recreation. A letter to the *Survey* imagined "a well-nigh universal ground-swell of feeling on the part of people of all sorts" in favor of daylight saving. "The daylight saving scheme has represented the annexation, as if from out of nowhere, of a new strip of life," continued the letter.[36] If clocks enforced simultaneity, unanimity, and unflagging obedience, here was an example of clock time creating a shared, apparently unanimous experience of wholesome, neighborly, renewing leisure.

Still, the idea of everyone experiencing the same relaxation at the same time every day gave credence to Veblen's charge about the mechanization of leisure. And the appearance of unanimity over daylight saving proved deceptive. Besides instituting daylight saving, the Calder bill also legalized the railroad standard time zones invented in 1883, and empowered the Interstate Commerce Commission to investigate the borders between the zones and alter them as it saw fit. The Commission's investigation disclosed a surprising diversity in America's timekeeping habits. State statutes legally governed the time in many instances, yet the Commission found those statutes widely ignored in places where railroads used a different time. Complicating matters further, in some of these places local communities used a still different time from the railroads. The Commission stressed that "it appears clearly from the record that there is need for a closer connection between the sun and the clock than has obtained in many parts of the country," and that moving the clock too far from the apparent time of the sun seemed both highly impractical and unlikely to succeed. Using these principles as its guide, the Commission systematically readjusted the zones, taking care to preserve the natural relationship to the sun and place the zone borders in sparsely populated areas as often as possible.[37] By respecting local timekeeping practices, and deviating as little as possible from established practice, the Commission managed to rule on standard time without awakening much opposition.

The same could not be said for daylight saving when the issue came before Congress again in 1919. In that troubled year, so scarred by violence and social change, the apparent consensus about daylight saving's virtues collapsed. After the armistice in November 1918, daylight saving lost the justification of wartime emergency. American farmers, who had kept largely silent on the idea during the war, began expressing their disapproval in no uncertain terms shortly after the new year. Farmers, it seemed, claimed that the new time forced them to rise in darkness and begin their chores too early. The cows refused to give milk, and the heavy morning dew made work in the fields impossible. "The farmer," the *Digest* summarized, "objects to doing his early chores in the dark merely that his city brother, who is sound asleep at the time, may enjoy a daylight motor ride in the evening."[38] As attempts to renew the daylight saving bill began, letters from disgruntled farm organizations began pouring into Washington. A typical petition, from the citizens of New Britain, Connecticut, echoed the claim that daylight saving inconvenienced the farmer, and added that "the time should be set by Nature, that is, by the sun, which is the best timepiece in the world."[39]

"The farmers," warned the *Times*, "are bombarding Congress with demands" for the repeal of daylight saving. "Write or wire immediately to [your congressman]," urged the New York Merchant's Association; "and urge your friends to also wire against the proposed repeal." "When we were informed," began an editorial in *Scientific American*, of a bill "aimed to kill the Daylight

Saving Act, we experienced a distinct shock." The bill had seemed so universally beneficial, and "we had always looked upon the farmer as one who delighted in early hours and who would love to have the rest of the world routed out of bed with the sun." To city dwellers, the farmer's hostility seemed inexplicable—did not daylight saving return society to closer harmony with nature, while saving fuel? Marcus Marks could only imagine that in fact the farmer had little to do with it; "although the sentiment against daylight saving seems to have come from farmers," Marks suggested, "it is really fostered by the large gas companies, which would benefit from repeal of the law."[40]

When Congress convened in June to discuss the repeal of daylight saving, the farmer's hostility occupied hour after hour of debate through the summer. Just why were farmers so upset? The congressional debate revealed the growing gap between rural and urban cultures, between patterns of life rooted in clock time and wage labor, and ways of living modeled on nature's example. It also anticipated the backlash against modernity that would later result in the Scopes Trial. Some farmers experienced genuine economic hardship as a result of daylight saving, and many more suffered serious inconvenience. But others resented having the arbitrary strictures of machine time imposed on their lives.

The two sides attacked each other in symbolic terms: urbanites, rural citizens claimed, lived off the farmer's ceaseless labor, yet had crust enough to demand more leisure time for their unproductive play. The urbanite lived in the thrall of the diabolical machine, while the farmer enjoyed the moral virtue of harmony with nature. Or, conversely, the farmer wallowed in ignorance and superstition, and trembled lest altering the clocks bring down the wrath of the almighty, while the city dweller tolerated and even welcomed modernity, innovation, and change. The farmer rhapsodized over nature, but selfishly allowed mere habits of tradition to deprive the toiling factory masses of the same sunshine and fresh air he praised so fulsomely. The debate over daylight saving in 1919 helped draw the boundaries between tradition and modernity, the conflict between inherited beliefs and scientific inquiry, that would characterize American intellectual life for decades to come.

The most common source of the farmer's complaint, the one most easily lampooned, was the morning dew. Under daylight saving, farmers protested, the dew evaporated an hour later than it had before, delaying the onset of field work and slowing the harvest. The farmer's hired help arrived at seven o'clock in the morning, as usual. But on daylight saving time seven by the clock meant 6:00 a.m. by the sun — the morning dew still lay heavily on the fields, and now the farmer had to figure out what to do with a whole crew of hired hands, standing idly, on the payroll, waiting for the dew to dry. We accepted this foolish arrangement during the war, rural voters wrote to Washington; now let us have a return to peacetime conditions.[41]

Urban congressmen found this objection preposterous — why not simply have the hired hands report an hour later than they did before daylight saving? If they began at eight, the dew would be gone, and if they then quit their eight-hour day at six instead of five the farmer lost no time. Many interpreted the farmer's apparent unwillingness to make this easy adjustment as pure stubbornness. Fiorello LaGuardia, then a representative from New York, called these arguments mere "camouflage." "Let us be frank about it," LaGuardia taunted his rural colleagues. "You are trying to repeal this law at the request of the employers of farm help, of the absent landlords . . . you are trying to get another hour of work out of [the laborer] at the end of the day." If farmers are so anxious to repeal all war measures, LaGuardia countered, "I ask how many of these gentlemen are ready to repeal war-time prohibition?"[42]

In fact, the situation was more complicated than LaGuardia

understood. The farmer might try changing the hours his help worked, but it seemed they adamantly refused to quit at the new time. For one thing, quitting at six instead of five deprived the farm laborer of the very benefits daylight saving was supposed to confer. Instead of gaining an hour of evening sunshine, the laborer lost it. Moreover, claimed an Indiana congressman, "the farm hand of today is rapidly assimilating the hours of labor in the industries, and, in rapidly increasing number, he quits with the blowing of the whistle in nearby towns."[43] When that customary quitting time came around, the hired help dropped their tools and raced off to meet their friends for baseball, movies, dancing, and the other activities of modern leisure. The problem came not from the farmer's hostility to innovation, or any superstitious dread of the dew, but from two opposed understandings of time. In the farm towns, clocks governed life and brought people together on schedule, but on the farm itself natural conditions refused to bend to the clock's rule.

City dwellers derived much hilarity from a second objection, that the dairy farmer's cows resolutely ignored the clock and refused to give milk by the new time. But here again two differing systems of time clashed, to the farmer's detriment. Suppose a dairy farmer traditionally milked his cows at five in the morning, and got his milk to the depot to be taken off for processing and sale by seven. Under daylight saving, he now milked the cows at six by the clock, too late in the morning to catch the milk train before it pulled away at seven o'clock. If he rose an hour earlier, he lost that much milk, if the cows gave milk at all. And of course he had been rising to his chores before dawn's light at four; now had to rise at 3:00 a.m., stumble around in the dark, and sell less milk, all so city dwellers could play at gardening after dinner. Citing his constituents' complete hostility to the measure, a Missouri congressman called daylight saving "the most mischievous and pestiferous experiment this country ever undertook."[44]

Farmers had enough trouble keeping their children "down on the farm" as it was, and daylight saving made farm life both less pleasant and more isolated than it had been. It deprived farm families of the modern pleasures and conveniences that were only just reaching rural America, for the farm family "is put out of joint with the situation in [the] local market town."[45] Nearby movies, lectures, meetings, dances, and other cultural and social events were over before farmers and their families could get to town. Farm wives rose with their husbands and worked in darkness, while their children either neglected vital chores or lost time in school.[46]

To make matters worse, some of America's most productive farm regions experienced far more drastic changes in daylight than the cities. The boundaries between standard time zones had been set to fall in rural areas as much as possible. A farm located at the extreme edge of a zone—like much of rural Ohio, for example—was already half an hour or more ahead of the "real" time of the sun, unless it ignored standard time. Daylight saving put some places nearly two hours ahead of the solar time; in some Ohio communities, the clocks showed "noon" at what had been one forty-five or even two in the afternoon by the sun. Such cases only compounded the problems already described.[47] These disadvantages combined to present farmers and their families with a considerable hardship, or at the very least with substantial and genuine reasons for wanting the bill repealed. But the relatively small economic and social hardship daylight saving engendered fails to explain its opponent's vivid rhetoric.

In their arguments against the bill, rural congressmen blamed it for the manifest ills of the age. It embodied the trend away from real production and toward idle consumption: the bill "is a pet of the professional class, the semileisure class, the man of the golf club and the amateur gardener, the sojourner at the suburban summer resort," charged a Minnesota representative, "who can all close their desks an hour earlier and hie them to an extra hour of play."[48] As a thoroughly non-productive, silly game played on the coiffed and manicured surface of perfectly good land, golf symbolized the utter decadence that underlay the daylight saving movement. The enemies of daylight saving scorned golf, again and again, as the wasteful indulgence of a parasitical class. By their honest sweat the farmers wrested the world's sustenance and ease from stubborn ground; instead of asking more sacrifice from the farmer, or laughing at his ignorance, rural politicians claimed, "you should rise up in your refinement and call him blessed."[49]

One of daylight saving's most outspoken opponents, Representative Edward King of Illinois, extended the attack on "refinement" to include progressivism generally. King saw daylight saving as the impractical child of the professional reformer and the university dreamer. During the war, King insisted, the farmer supported all sorts of boneheaded "scientific" measures for " 'economy,' 'efficiency,' 'coordination' " in good faith, even to the outrage of commonsense. But now King lambasted "the impractical, the Utopians, the theorists, political astrologers, medicine men, and advance agents of the millennium" who, he insisted, constituted daylight saving's main support. A year earlier one of King's colleagues had laid daylight saving's invention to "an uplift magazine"; now King charged its continuance to "that intangible professorial influence which we all recognize as occupying a sort of dictatorial attitude during the war and whose decrees we have so often observed and felt." Desirous of "applying the pedagogical whip to the backs of what they deemed the unruly and uneducated people," "the Professariate" have forgotten in "their theoretical air navigation both God and nature."[50]

Doubtless much of the sentiment against daylight saving reflected a general weariness with the pretensions of the "Professariate," and with Wilsonian Progressive crusading in particular. Wilson twice vetoed the repeal of daylight saving after the appropriate majorities of both houses had approved it, in pronouncements thick with Progressive pieties. His first veto message lauded the "competent men" whose "careful study of industrial conditions" had led to the bill's formulation, and his second stressed the "immediate and pressing need" for "increased and increasing production." The bill "ministers to economy and efficiency," Wilson suggested, but his congressional opponents would have none of it.[51] "Now that the whole world has been made safe for democracy," remarked a member from Illinois sarcastically, "perhaps it is appropriate that the President should insist that every day in this country be made safe for the joy rider and for the patron of the golf links and tennis courts."[52] Members of Congress frequently suggested that perhaps the President should spend more time at home, listening to the people, and less time in Europe saving the world.

But beneath all these objections there lurked a fundamental resistance to the clock's authority, and to the mechanization of society. The "Professariate," Congressman King had charged, have forgotten in their theories both God and nature. Again and again, rural Americans and their representatives pledged their allegiance to natural time, God's order, and "nature's timepiece," the sun. Senator Oscar Underwood of Alabama spoke for many when he insisted that "time has been fixed for ages by the movement of the sun." Now that the war is over, he asked, "let us stand by the custom that generation after generation has adopted." Daylight saving interferes, added an Iowan, "with the natural order of things as regulated by the rising and setting of the sun." Come out to Iowa, Congressman Burton Sweet invited his urban colleagues, "and really live as you should live," not in violation of "the mandates and decrees of the Eternal God." Representative Ezekiel Candler of Mississippi delivered perhaps the most stirring peroration on behalf of the sun when he angrily insisted that "the rising of the sun and the going down thereof fixes the time." "God's time is true," Candler rang out; "man made time is false." "Let us repeal this law and have the clocks proclaim God's time and tell the truth," he thundered to his colleagues' applause. "Truth is always best. It is mighty and should always prevail."[53]

Their urban opponents found these claims patently ridiculous. Since 1883, they pointed out, Americans had been using four standard time zones based only loosely on the sun's position, with no ill effects, no celestial wrath. But these countercharges missed the point. For one thing, congressional investigation had only recently disclosed that where people found the standard clock time too far removed from the sun, they simply ignored it; and indeed as Chapter Three discussed, their decision to ignore standard time had the backing of law in many states. But more importantly, rural congressmen argued not for perfect fidelity with the sun, but rather for a life regulated by natural principles, in which clocks served as useful approximations of the time rather than time itself. When daylight saving threw their lives out of synch with nearby towns, they interpreted the move as slavery to clocks and industrial clock time.

Declarations of principle like Congressman Candler's brought only scorn from the partisans of daylight saving, for whom such phrases sang of a thick-minded, foolish obstinance. By couching their arguments in religious terms, farmers made it easy for their opponents to tar them with the brush of ignorance, backwardness, and reaction. One typical letter to Congress, for example, charged agitation against daylight saving to "a few mossbacks, hayseeds and night riders" making a bigger noise than their number warranted.[54] In an illustration in the *Literary Digest,* a bearded farmer shook his fist at a departing automobile, which in its speeding haste had knocked the cans of milk he carried out of his wagon. At the wheel of the car, Father Time left a cloud of dust as he stepped on the gas. "UPSETTING EFFECT OF FATHER TIME'S NEW BURST OF SPEED," read the caption. A Philadelphia editor attacked the "obstinate pigs and cows who dictated to Congress" on daylight saving.[55] These jibes fed the desire to seem modern and "up to date." But in assigning the movement for repeal to what they imagined an ignorant, hidebound minority, daylight saving's partisans ignored the class and social biases built into the daylight saving scheme.

Though its supporters claimed universal benefits for the measure, in fact daylight saving best served certain limited segments of society—middle-class, urban wage earners in the Northeast especially. It penalized groups more dependent on natural rhythms—farmers, obviously, but also housewives, children, and less skilled workers in low-status jobs—as well as people living near the boundaries of time zones. During the debate over daylight saving most of the attention focused on the farmers, whose legendary conservatism supposedly explained their resistance to modern, rational innovations. But the measure might never have been repealed had only farmers opposed it. The fatal blow to daylight saving came at the hands of industrial workers, and particularly the American Federation of Labor.

For certain groups of workers, daylight saving only worked an inconvenience and depressed the spirits. For example, workers unable to afford a house within easy commuting distance of their job, then as now, had to rise that much earlier than their more fortunate co-workers. Most of the year they rose in darkness; only in summer did they have light to dress by.[56] Daylight saving deprived them of that morning sun. Representative King of Illinois pointed especially to the coal miners of his state, who had voted, along with the Illinois State Labor Federation, for the repeal of daylight saving. The situation in Illinois was made doubly unpleasant by the state's location in the far eastern section of the central time zone. Here were men engaged in difficult, arduous work, vital to the nation, in nearly constant darkness. Should they be deprived of morning sun just so city dwellers could play golf?[57]

Starting early often meant finishing early, and one of the few benefits of such jobs had been the extra daylight after work. But with the new time, the blessing became a curse as the extra daylight, and the noises of summer recreation, often intruded far into the hours for sleep. In parts of the Midwest especially, the sun that once set at nine o'clock now set at ten-thirty or even later, making it all the more difficult for parents to get children to bed, and to get to sleep themselves. A second problem, one we are inclined to take for granted in the air-conditioned present, stemmed from the resulting heat. Before daylight saving, workers made their way home in the waning heat of day, and enjoyed the cool evening hours sitting on the porch or sleeping; but under the new time an extra hour of hot sunlight made working conditions, and the evening, all the more unpleasant. "These are trifles, you will say," wrote "a workingwoman" to *The New York Times*. "Yes, but trifling things like flies and mosquitoes are a nuisance. Why continue?" During the war we were all glad to make small sacrifices, but now we workers too want an end to daylight saving — "it is not all the farmers, as you seem to think," she added.[58]

Labor leaders like Samuel Gompers pronounced themselves all in favor of daylight saving, and their testimony had helped pass the bill in the first place. But at the AFL's June 1919 convention, the delegates soundly defeated a measure protesting the repeal of daylight saving. Prominent speakers against the resolution included John L. Lewis of the United Mine Workers, whose members resented having to rise as early as three-thirty in the morning, sun time, and trudge to work in gloomy darkness. The Convention's actions gave credence to the farmer's charge that daylight saving in fact favored mostly idlers and the wealthy classes. There is no way of judging the actual extent of workers' opposition to daylight saving time. But the AFL's rejection of the bill, coupled with other scattered letters and resolutions by lesser labor organizations, hints very strongly of opposition to daylight saving among the least skilled, least organized, and least represented members of the working class.[59]

These objections to the measure point not to any great harm worked by daylight saving, but rather to the inequity, in 1919, of trying to regulate all social affairs by one clock time. There seems little doubt that daylight saving worked a positive good for most, if not all, factory and office workers in the cities. But benefits for one segment of society equaled losses for another, and the enormous differences in patterns of living between different groups, across so vast an expanse of land as the United States, made the clock-changing law impractical at best. Over twenty-eight bills for the repeal of daylight saving were introduced in the summer of 1919. Two surviving bills reached President Wilson's desk, only to be vetoed each time. Congress finally overrode Wilson's second veto on August 21, 1919, ending national daylight saving until the next world war.

Almost immediately, Marks and the United States Chamber of Commerce began lobbying for "daylight saving by local ordinance." The original Daylight Saving Act of 1918 had legalized standard time zones and empowered the Interstate Commerce Commission to establish the time regulating interstate commerce. The repeal movement focused only on national daylight saving, and nothing in the law seemed to prevent cities, states, and local communities from adopting any time they wished, as they had before the war.[60] Massachusetts adopted a statewide daylight saving law in 1925, prompting the Massachusetts State Grange to bring suit against the measure. The case reached the United States Supreme Court, and Justice Oliver Wendell Holmes, a year later. Over forty years before, as a Massachusetts Supreme Court Justice, Holmes had offered the nation's first ruling in favor of standard time. Now Holmes ruled that individual cities and towns could legally adopt daylight saving if the difference from standard time amounted to no more than one hour. Thereafter, most major cities and some states adopted daylight saving in the summer months, simply ignoring the hour difference between the time their environs and the railroads used.[61]

Which brings us to an intriguing question: Who stood to gain, or lose, the most by saving daylight? The measure's recreational benefits seem the most obvious today, and indeed they figured prominently in every account of the law's virtues. Many of those businesses involved in recreation, like baseball leagues, for example, stood to gain dramatically from extended daylight. Shortly before the passage of the Calder bill in 1918, President Ban Johnson of the American Baseball League spoke enthusiastically of plans to begin ball games an hour later under the new time. Previously (before artificial lights), games started at three-thirty, leaving many office workers no time to get to the ballpark. Under the new time, sunsets came a hour later. Beginning games at four-thirty gave fans more time to get to their seats, raising attendance and profits. Here seemed a perfect example of the charms of saving daylight, especially for the wallets of baseball leagues and team owners.[62]

But the plan to change hours collapsed under pressure from Washington. "Slackers," declared Charles Lathrop Pack of the War Garden Commission—"Slackers of the worst type is the brand placed upon baseball team owners or managers who plan to move down the scheduled times of starting games." President Wilson, Pack recalled, had named food production for the war effort a top priority, and to waste the increased daylight at the ballpark "violated the spirit of the law." Changing the starting times "will take thousands of hours of time from gardens," and the daylight saving law, Pack preached, "was not intended to give extra hours of recreation." "I hope," he added, "this is not the attitude of those in control of the greatest of national games." In the climate of war, such appeals touched a weak spot in the nation's psyche. That same day President Tener of the National League responded contritely that no, "he had not heard of any contemplated changes in the time of games."[63]

For Pack and the War Garden Commission, thousands of baseball fans flinging down the sports page and heading out to buy garden tools sounded like money. Despite its official-sounding name, the War Garden Commission was in fact a lobbying organization for the makers of garden products—tools, seeds, fertilizers, canning and preserving equipment, including jars, cans, and rubber ring seals—who stood to gain dramatically from any increase in wartime gardening. Pack issued pamphlets and newspaper advertisements insisting that "the preservation of vegetables and fruits . . . is a patriotic duty and a national war time need." Accompanying the message of conservation came a list of manufacturers of garden products. Masking itself as a government agency, Pack's organization used daylight saving to raise sales, and used patriotism to head off the competition.[64] In this case, one recreational business—baseball—lost out to another—the makers of gardening and canning equipment.

By 1919, the United States had developed substantial and important industries based on leisure time. Chambers of commerce and boards of trade were daylight saving's most enthusiastic supporters by far, and though it has proven difficult to establish any direct link between the organizations that supported daylight saving and increased profits, much evidence suggests a connection. For example, A. Lincoln Filene, one of daylight saving's strongest supporters, owned Boston's largest department store chain. Women customers especially may have been more comfortable shopping under the new time, because, as the New York State Republican Women's Association put it, the "law safeguards the young woman returning from her business before darkness has fallen."[65] If women felt safer walking home in the light, then daylight saving gave them an hour more time for shopping in Filene's stores.

Daylight saving particularly threatened motion picture companies, and especially theater owners, who saw attendance fall drastically on pleasant summer evenings. A scrap of silent film title in this writer's possession, from approximately 1915–16, declares: "If California adopts daylight saving, she will tell the world she hasn't enough sunshine. Say NO to daylight saving."[66] California toyed with adopting statewide daylight saving in 1916. The piece of film was apparently part of a counterattack co-sponsored by the movie industry and real estate developers, who promoted California as the land of warm, boundless sunshine and oranges. Later, in 1930, Fox West Coast Theaters of Los Angeles led a second fight against daylight saving for California, which its spokesman claimed brought "unlimited possibilities for evil" to the movie industry. Fox pledged $50,000 to stopping such a measure.[67]

More recently another lobbying organization, the Daylight Saving Time Coalition, managed to move back the starting date of daylight saving, from the last Sunday in April to the first. James Benfield of that organization works on behalf of the makers of sporting equipment, charcoal briquets, insect repellants, and other accouterments and necessities of leisure time, including Seven-Eleven stores, which do more business by daylight, and candy companies, who maintain that daylight saving on Halloween night keeps small children out longer and increases candy sales.[68] For years, Benfield complained, politicians from certain states — Idaho, for example — had refused to see the benefits of daylight saving. When he discovered that fast-food franchises do more business under daylight saving — "up to eight hundred dollars per unit per day" — he called the leading franchises to ask where they got their raw supplies. Told that Idaho supplied McDonald's with the bulk of its potatoes, Benfield relayed the story to that state's representatives in Congress, who promptly reversed themselves and henceforth went along with daylight saving like a burger with fries.[69]

Changing the clocks by law, then as now, inevitably favored certain groups at the expense of others.

Daylight saving most benefited those segments of society that were strictly governed by clocks and clock time. It gave them extra daylight for recreation in a period that saw a virtual explosion in recreation and leisure time pursuits. The same objective — more daylight after work — could have been more easily, or at least more fairly, achieved by the simple expedient of rising earlier, or changing work schedules. But as one editor suggested, "so completely have we become slaves of the clock that when we think it desirable to get up an hour earlier we could not think of any better way than to pass laws setting the clocks an hour ahead of time."[70]

The apparent unity daylight saving encouraged was one of its most attractive points: it gave the impression of a society united in relaxation and cooperative leisure, united in rejecting, to some extent, the very principles of efficiency, industriousness, and temporal thrift that had given the clock such power in the first place. "100,000,000 people changed by an hour their time of getting up in the evening and going to bed at night — and of everything they do between," marveled the *Times*; "when one remembers the number of activities, great and small, that were affected . . . the disproportion between the means employed and the results accomplished is nothing less than astounding," a "revolution without one tremor."[71] The word "revolution" was an interesting choice. Daylight saving readjusted the predominant governing mechanism of urban society — the clock — to suit that society's new desire for greater leisure. It represented a kind of triumph over time, a revolution both in the sense of control over time that it fostered, and in its pure celebration of pleasure and fun over workday dreariness.

But daylight saving contained a paradox, in that it used the key tool of a mechanized society — the clock — to create a sense of freedom from mechanization and clock time. The great thing about daylight saving, according to the *Times*, was that everyone apparently experienced the same feeling of freedom and relaxation at the same time. Daylight saving mechanized leisure in the pursuit of freedom from mechanization.

The resistance to daylight saving, and the law's eventual repeal after two presidential vetoes, points to stubborn and lingering hostility to clock authority, to a refusal to abandon natural models for organizing life's work. Rural Americans used clock time to symbolize a wide range of changes brought about by industrialization—factory labor and mechanization, the political and economic dominance of the city, the growth of nightlife and leisure, a perceived decline in the work ethic, the "scientific" reformation of society by academic experts. If we might fault them for willful ignorance, we might also applaud their acute choice of symbols. In daylight saving, they found an unparalleled example of the falsity, the self-delusion, the distance from natural imperatives they perceived in a society regulated by mechanical timekeepers.

The daylight saving debate offers only one example of how "modernity" changed time. It seemed in a small way to make men and women godlike—as Joshua commanded the sun to stand still, so the modern nation stopped its clocks to suit its needs. But American technology had always inspired wonder at how the United States, like a child imitating its parents, seemed to constantly encroach on the Almighty's prerogatives. "God said, 'let there be light,' and there was light," reminded Senator Orville Platt of Connecticut in an 1891 speech on "Invention and Advancement." Think, he continued, of the electric light. "You will feel it scarcely irreverent to exclaim 'And man said, 'let there be light,' and there was light."[72]

But did human emulation of God drag God down to earth?

Did advancement lead only to arrogance and immorality? In a 1919 essay on "the psychology of daylight saving," an Iowa professor pointed out that with the electric light, "there has come about a displacement between the daylight day, and the human day," so that we rise well after daylight and stay up long into darkness. This displacement of the "natural" times of work and rest is now "going on faster than ever," he wrote. "Every year it seems a little harder to go to bed or to get up at the old accustomed time. Automobiles, unlike the horse, travel as well by night as by day and keep increasingly late hours. College and high school students study later at night or engage in social activities . . . more and more industries continue through the night." "Without doubt the number of people who sleep to noon is constantly increasing," he warned his readers, and "darkness is a cover for every evil thing." How far shall we go? he asked. "Will day and night finally be wholly transposed?"[73] In so many ways, "modernity" upset the traditional relationship between time and the natural world.

Modern science called old authorities for time into question. After 1920, for example, astronomers began using clocks accurate enough to detect variations in the earth's rotation. These modern clocks regularly revealed the passage of the days, once the sine qua non of steadiness and sober reliability, as little better than an uncertain and capricious wobble. Even worse, since the war newspaper editors and scientific magazines had kept up a steady babble about time's relationship to the speed of light and the observer's point of view, a whole new physics of time.[74] Each development questioned the authorities for time.

Modernity altered not just the authority, but the *experience* of time as well. In *Middletown* (1929), the pioneering sociological study of the American Midwest, Robert and Helen Lynd recorded the automobile's effects on life in Muncie, Indiana. Owning a car expanded the owner's spatial and temporal universe. But it also destroyed old customs and altered family relations. Instead of staying home, or socializing with neighbors, friends, and family, auto-mad Americans dashed off to distant pleasures. The Lynds described the automobile "blasting its way through such accustomed and unquestioned dicta as 'Rain or shine, I never miss a Sunday morning at church;' 'A high school boy does not need much spending money'; 'I don't need exercise, walking to the office keeps me fit'; 'I wouldn't think of moving out of town and being so far from my friends'; 'Parents ought to know where their children are.'" They observed a Sunday School class. "Can you think of any temptations that we have today that Jesus didn't have?" asked the teacher. One boy blurted out "Speed!"[75] Not just speed in automobiles, but speed in daily life—fast music, fast dancing, fast living, late hours. Modernity shattered notions of time and space, and disrupted traditional relations between the individual and society.

In the 1920s, opposition to modernity took many forms, including Prohibition and movie censorship, which both involved attempts to control leisure time. But no single attack on modernity attracted as much attention as the fundamentalist movement and its crusade against evolution. Historians often treat American fundamentalism as a reactionary movement of the disaffected, a dangerous trend born in ignorance. Educated urban elites have always found fundamentalism—and its substantial popularity—threatening. H. L. Mencken unfortunately set the tone for much subsequent academic discussion: "Heave an egg out a Pullman window," Mencken sneered in 1926, "and you will hit a fundamentalist anywhere in the United States today." According to historians, fundamentalists suffered from "status anxiety"; they perceived themselves losing power and reacted irrationally, denying scientific progress and imagining vast modernist conspiracies in "the paranoid style."[76] Doubtless fundamentalism offers plenty of ammunition for such attacks. But American fundamentalism also addressed a widespread and genuine sense of unease about modernity and modern time, an unease that deserves to be taken seriously.

The fundamentalist movement had its roots in Millenarianism, or the belief in and study of Christ's return. Millenarians scrutinized biblical chronology to chart the present status of God's prophecy—they used the Bible as a record of time to predict the future. In contrast to the buoyant confidence and optimism typical of the antebellum years, most American Millenarians looked around them and saw evidence of immorality and approaching collapse. The extraordinary popularity of Millenarian thinking—Mormons, Seventh-Day Adventists, and Sabbatarian crusaders all drew from Millenarian ideas—constituted a popular critique of secular industrial culture.[77]

Nineteenth-century American Protestant thinkers had also drawn heavily on a Scottish intellectual tradition, "Common Sense Realism." "Common Sense" philosophy insisted that the evidence of the senses was reliable, and that ordinary people could be trusted to know and grasp the truth. It depended on an interrelation between science, the Bible, and the natural world. "If God is the Creator of the Universe," insisted a Yale theologian in 1852, then "the study of the creation [must] be interlinked at every point with the study of the Creator, and thus become, to the devout mind, the study of theology." When *McGuffey's Readers* encouraged students to study nature for models of Christian good behavior, they reflected common-sense realism.[78]

But when modern science studied the world, it reached conclusions that disputed the Bible account of creation. Geology, for example, turned up evidence of the earth's vast, almost incomprehensible age, and fossil records of animals and plants that the Bible never mentioned. Most American theologians managed to reconcile modern science with Christianity by abandoning their belief in the literal truth of Bible accounts. As the challenges of modernism continued, however, more conservative "fundamentalist" theologians began emphasizing the Bible's "inneracy," or literal truth. When science contradicted the Bible, fundamentalists presumed, it had moved away from religious truth and the evident "common sense" gained from observing the world in everyday life.[79]

By the twentieth century this emphasis on the literal truth of the Bible had become a crusade. One Georgia legislator recalled common-sense philosophy when he directed his constituents to "read the Bible. It teaches you how to act. Read the hymn book. It contains the finest poetry ever written. Read the almanac. It shows you how to figure out what the weather will be. There isn't another book that is necessary for anyone to read."[80] His statement linked nature to the Bible in terms of practical daily affairs, recalling schoolbook lessons then close to a century old. The Bible and the almanac—together the two books spelled out the nature of time and the terms for scheduling life.

In the twentieth century, "fundamentalism" came to serve as a linguistic umbrella sheltering a wide range of anxieties and fears about the direction of American society. American cultural "fundamentalists" worried particularly that new ways of experiencing time threatened the foundations of morality—late hours, necking in movie theaters, Sunday drives instead of churchgoing. And the most pressing temporal challenge to tradition came from the theory of evolution. Much of the fundamentalist opposition to evolution is familiar—it seemed to degrade human beings, to bring men and women down to the level of apes. But Darwin's theory also subverted traditional notions about time in profound and crucial ways.

Evolution upset the "temporality" of Genesis. Claiming that higher animals had evolved from lower did more than just deny God's act of creation. It denied Genesis' structuring of time—the creation week from the first day through the seventh. If organisms evolved over long ages of years, instead of appearing at once on successive days, then the seven-day cycle of creation made no sense. And it was those seven days that gave Judeo-Christian religious practice its temporal structure. The "clock-like rhythm of the weekly Sabbath" was deeply rooted in American culture, and Sabbath observance helped sanctify the week's labors and keep the family on an even moral keel. Take away the creation week, and one day was just like another; you removed all justification for Sabbath observance, and desecularized the framework, the schedule, that organized daily life. The weekly ritual became arbitrary, empty motion devoid of any meaning other than social convenience.[81]

Darwin's theory depended on the earth's great age: evolution could only have occurred over millions of years. Starting in the eighteenth century, geologists had found irrefutable proof of "deep time"—proof that the earth's history stretched back almost beyond comprehension. Evolution linked the "deep time" of geology to a narrative of human development.[82] And once again Darwin's history of man threatened the temporality of events in the Bible.

In Genesis, God established "the day" as the measure of time. The creation story linked the present to the very beginning in a series of recognizable, familiar increments. The Bible gives a historical record, the narrative account of humanity's progress. If the single day was taken as a stable unit of measure, then the Bible could serve as an accurate record of *all* time. Seventeenth-century European scholars, including Isaac Newton, had made strenuous efforts to calculate the earth's age from Scripture. In 1658, Bishop James Ussher of Ireland used the record of generations to make his famous estimation: the earth was created on October 23, 4004 B.C. American Protestant Bibles almost routinely offered the bishop's dates as footnotes or supplementary material.

For ordinary readers, Ussher's chronology connected biblical events to the fundamental unit of "natural" time—the day—and gave Bible stories a tangible connection to the present. A "day" in 1920 was just like a day Jesus lived in Jerusalem, Moses passed in the desert, and Adam saw at creation. To study nature was to study creation; to work six days and rest on Sunday was to relive, in a small way, the cycle of Genesis. The passage of days, one after another, from past to present to future offered confirmation of God's prophecy unfolding. But Darwin's alternative chronology stood Ussher on his head. It severed the connection between nature, human history, and the Bible, knocking one more brick from the foundation of biblical authority.

Evolution robbed time's passage of its meaning, or rather, it substituted new meanings for old. Nineteenth-century "social Darwinists" like Herbert Spencer related Darwin's theories to social class on a crude model of "survival of the fittest." In this view the rich got rich because they were more fit than the poor, while the poor, conversely, suffered because of inferior genes. Most such misreadings of evolution hinged on enthusiasm for progress. It only took a bit of intellectual carpentry, for example, to make evolution dovetail with notions of efficiency and industrial advancement. If Darwin's laws of evolution produced better and better organisms, then the progress of invention must be evolving ever better and more efficient mechanisms, better and more efficient ruling classes. In this view, the process of getting rich "measures the amount of efficient management that has come into the world and the waste that has been eliminated." Late nineteenth-century Americans, especially those who had benefited most from industrialization, hailed Spencer as a genius, the prophet of a new order.[83]

But Spencerian evolution could also seem cruel, immoral, even anti-democratic. Instead of hailing "the people" as the backbone of the republic, as common-sense philosophy had, it branded ordinary, plain people evolutionary losers by their very lack of worldly success. These were, of course, misinterpretations: natural selection never produces "better" organisms, only different specializations for different environments. But much of scientific management's appeal stemmed from seeing evolution as a mechanical process of constant improvement, and relating that vision to society. In simplistic interpretations of Darwin's theory, "time" meant "the machine-like progress of efficiency," not the revelation of God's will.

Linking evolution to technological and social progress made it a hard theory to argue against, as long as progress seemed benign and steady. World War I, however, shattered confidence in progress and the benevolence of social evolution, for instead of vindicating dreams of a more enlightened future it presented a world of nightmarish, mechanized death. American fundamentalists saw the war not only demonstrating the failures of modernity but also vindicating their hostility to Darwin, because the war denied that the secular time of evolution led to a better world.[84]

By the 1920s, evolution had emerged as the fundamentalists' best target in the fight against modernity. Linked to modern science in a general way, evolution supported most of the false gods that corroded morality and tradition. Southern legislatures especially began introducing measures designed to ban or restrict its teaching. These laws offered fundamentalists both an effective symbolic action against modernism and national media attention. The most dramatic confrontation between fundamentalism and the modernist impulse occurred in 1925, on a hot July day in Dayton, Tennessee, when Clarence Darrow brought William Jennings Bryan to the stand in the notorious Scopes "monkey trial." At the Scopes Trial, fundamentalism and modernity clashed over the issue of time.

Tennessee had passed a "monkey bill" earlier in that year, prohibiting the teaching of evolution in state schools. Scopes, a Dayton high school teacher, freely admitted that he had violated the statute and agreed with the American Civil Liberties Union to make his a test case. When word spread that Clarence Darrow would appear in Scopes's defense, and William Jennings Bryan assist for the prosecution, the trial quickly turned into a national obsession. The two men symbolized the opposing forces of tradition and change.

William Jennings Bryan burst on the American scene at the 1896 Democratic Convention, where the young Nebraskan had electrified his party with his famous "cross of gold" speech. The speech's impassioned defense of the poor and downtrodden over the wealthy and privileged earned Bryan a reputation for oratory, and when the Populist Party endorsed him in 1896, it cemented his association with rural discontent and hostility to industrial capitalist excesses. Three losses in the race for President failed to tarnish his reputation as the people's champion, and throughout his career he remained true to his nickname, "the great commoner." Bryan never really believed in a strictly literal interpretation of the Bible, as the Scopes Trial would reveal to his detriment. But this profoundly religious man shared both the fundamentalists' unease about the direction of American culture and their hostility to modern science. "It is better to trust in the Rock of Ages than to know the ages of rocks," he often said; "it is better for one to know that he is close to the heavenly Father than to know how far the stars in the heavens are apart." By 1925 he had emerged as America's most prominent critic of modernism.[85]

His opponent at Dayton, Clarence Darrow, also grew up in farm country, in rural Ohio. But there the resemblance largely stopped. Darrow's father, something of a village eccentric, taught him to doubt and question accepted wisdom almost as a reflex. He became a lawyer, he said, mostly because he wanted to avoid working for a living. A religious agnostic, Darrow made his name defending unpopular causes, especially those that offended the religious sensibilities he grew up challenging. He defended Eugene Debs in the Pullman strike, Big Bill Haywood and the IWW, Chicago murderers Leopold and Loeb, as well as Hinky Dink Kenna, the Chicago political boss. In his commitment to free speech and cosmopolitan inquiry he at least equaled Bryan's passion for the average man, and like Bryan he fancied himself the underdog's champion. But where Bryan clung to the values of his rural past, Darrow saw those values as shackles to cast off. By the time of the Scopes Trial, he was America's most famous—and notorious—trial lawyer, spokesman for marginal causes, and darling of the H. L. Mencken/*American Mercury* set.[86]

Even Mencken left his accustomed haunts for the Scopes Trial, drawn by the prospect of a clash between Darrow and Bryan. He was not alone; the humble courthouse in Dayton, Tennessee, drew journalists from around the country, including technicians struggling with motion picture and radio equipment. A national audience awaited the outcome of the "monkey trial," and Dayton quickly took on the atmosphere of a circus. For weeks the surrounding hills echoed with the sounds of Holy Rollers, while in town, a motley group of itinerant preachers, hot dog and soda vendors, trained chimps, would-be prophets of the apocalypse, and cranks of all sorts vied for media attention. "The thing is truly fabulous," Mencken crowed. "I have stored up enough material to last me twenty years."[87]

Once the trial got under way, Scopes's lawyers bridled under the rule of Judge John T. Raulston, favorite son of Gizzards Cove, Tennessee. Raulston, stopping frequently to pose for cameramen, barely bothered to hide his empathy for Bryan, rural America, and the fundamentalist cause. He may have even set his watch by local sun time.[88] A banner emblazoned "Read Your Bible" hung over his courthouse door, and he began the trial by quoting Genesis, his voice carried by loudspeaker to the crowds gathered on the lawn outside. The first days of the trial saw considerable wrangling over whether or not the judge would allow scientific testimony. When Raulston ruled it inadmissible, it seemed to leave the defense without a case. But then Darrow called Bryan to the witness stand. By this time, driven by the stifling heat, the court had removed itself to the courthouse lawn, where a large crowd listened intently.

Again and again, relentlessly and even cruelly, Darrow challenged Bryan's belief in the literal truth of the Bible. Most of his questions concerned time and time measurement. Darrow asked Bryan if he really believed that Joshua had made the sun stand still. Didn't the earth in fact revolve around the sun? Bryan dodged that question, but Darrow quickly began asking about the date of the great flood—2348 B.C., by Bishop Ussher's calculation. How did Ussher arrive at his figure? Darrow handed Bryan a Bible giving Ussher's dates for major events at the top of each page. "How old is the world?" Darrow asked again; "how long ago was the great flood?" Someone produced a pencil and paper, and the two of them, absurdly, began doing addition and subtraction.

Ironically, the two men were engaged in an old American exercise. Nineteenth-century schoolbooks routinely used the Bible for arithmetic lessons—we know a day is 24 hours, and a year 365 days, primers reminded, so careful calculation should give us "the interval from the Creation to the birth of Christ," or tell us "how many seconds have passed since the creation of the world."[89] Both Bryan and Darrow had probably tackled these or very similar questions in childhood, as part of their math exercises. In their youth everyday experience—literally *every day* experience—combined with mathematics to confirm the Bible's truth.

But modern standardized time and science undid that relationship. Darrow quoted from geologists who estimated the earth to be hundreds of millions of years old, and cited proof of ancient civilizations far in advance of Ussher's dates. When at one point in the trial Bishop Ussher was quoted placing creation, with lunatic precision, at "9 a.m., on October 23, 4004 B.C.," some audience wag had shouted out, "eastern standard time!"[90] The point, of course, was that deprived of its religious origins, "time" seemed to have lost its definition and become arbitrary, meaningless.

Or rather, that time had gained a new meaning, more suitable to industrial society. Where the Bible defined time as the solar day, industrial society defined time as whatever it pleased, so long as all agreed to use it.

So when Darrow asked Bryan, "Do you really think the earth was made in six days?" he knew the charge the answer carried. Did the evidence of the senses, the natural world, offer a valid model for comprehending time and organizing society? Or was time simply an arbitrary construct, variable at will? Bryan's answer reportedly shocked his supporters in the audience: "Not six days of twenty-four hours," he replied. "Doesn't the Bible say so?" Darrow continued. "No sir," retorted Bryan. The response brought gasps from the crowd, and a cry of "What does he want to say that for?"[91] Darrow pressed Bryan further, till he confessed that the "days" of creation might have been periods of indefinite length. Bryan had never wholly believed in the Bible's literal truth. But he had appointed himself the defender of fundamentalism, and his dismayed supporters counted the admission a victory for the other side. If you questioned the day as a measure of time, then you questioned the natural world's use as the pattern for living—questioned, in other words, the ways people organized their time, their lives, and the things they chose to believe.

Bryan and Darrow's confrontation has passed into national folklore; the play *Inherit the Wind*, later made into a movie starring Spencer Tracy as Darrow and Fredric March as Bryan, depicted Bryan as a foolish, windy, and defeated old man, and interpreted his loss as a triumph of learning and intellectual freedom over bigoted repression.[92] Darrow did make Bryan look very bad—he proved himself poorly educated, evasive, and unable to respond to fairly standard questions. The trial came to symbolize a conflict between urban and rural America, between "enlightenment" of the H. L. Mencken variety and what Menckenites thereafter permanently labeled "ignorance."

But fundamentalists, rather than being simply "ignorant" or anti-intellectual, in fact stood fast by an older conception of science and the world. They judged "the standards of a later scientific revolution"—Einstein's, Darwin's—"by the standards of the first—the revolution of Bacon and Newton."[93] The critics of daylight saving understood time in the same terms—the terms of Newton's mechanistic universe. Like Bryan, they took the rationality, the perfection of natural movements as an article of faith, and saw nature's timekeepers as harmony with those principles.

Bryan died shortly after the trial. But before he died, he wrote a lengthy response to Darrow and the evolutionists. Though he clearly misunderstood most of Darwin's theory, he left some eloquent statements of why evolution struck him as a profoundly immoral doctrine. Evolution, he claimed, "discourages those who labor for the improvement of man's condition." Instead of giving them a tangible, meaningful connection with the past, "it chills their enthusiasm by substituting eons for years. It obscures all beginnings in the mists of endless ages." Evolution, in other words, severed the link between time and God established in the Bible. It replaced the unfolding of God's will from day to day with a purely mechanical, "cold and heartless process, beginning with time and ending in eternity." Like a machine, evolution operated on imperatives of efficiency, according to an arbitrary framework of time, indifferent to human needs.[94]

Bryan joined his hostility to evolution with an attack on modern science and industrial machinery. "Science . . . can perfect machinery," he charged, "but it adds no moral restraints to protect society from the misuse of the machine. It can also build gigantic intellectual ships, but it constructs no moral rudders for the control of storm-tossed human vessels." Bryan perceived the theory of evolution as an intellectual by-product of industry and industrial time. It replaced the natural cycles of the Bible, unfolding with a purpose toward the resolution of God's will, with a hostile, mechanistic process, a savage competition occurring over a mindless span of empty ages.

Bryan tied evolution specifically to the modern machinery of warfare—submarines, airplanes—through which "science has proven itself an evil genius." We read almost every day, he noted, of newer and more terrible weapons of destruction, and industrial evolution seems to carry us toward disaster.[95] His point is still relevant today—who never wonders if the dangers of technological progress don't outweigh the benefits? Evolution, of course, is not responsible for nuclear proliferation or global pollution. Like Bryan, American fundamentalists responded not to Darwin's writings per se, but to the context they confronted them in—scientific management, factory labor, the decline of American farm culture. Bryan and the fundamentalists linked belief in evolution to a society governed by mechanical principles, by unmoral machines and heartless standards of efficiency. By what authority, they asked, do we govern progress? By what authorities do we find the meaning of our time?

Epilogue

I n the one hundred years between Eli Terry's Connecticut
clock and the Scopes Trial, Americans evolved a new sense
of time based on machines. In 1825, time had come from nature
and God; women and men lived *in* time, just as they lived in nature
and the world, and nature served as the example for organizing
time. But industrial progress demanded a uniform medium of
exchange, a way of rationalizing and ordering interstate commerce
and communication. By 1925, most people got their time from
clocks and watches. Arbitrary, uniform, easily distributable, this
new time kept industry and commerce running smoothly, but it
drove a wedge between men and women and the natural world,
forcing them to divide their lives between "work time" and
"leisure."

Standardized, machine-based time also established new priorities for organizing society, new models for patterning daily life.
Clock and watch advertising insisted on a merger of identity between timepieces and their owners, in which clocks served as the

standards for personal conduct and good industrial citizens could be measured by the extent of their fidelity to their timekeepers. Standardized, arbitrary, divisible time made it possible to restructure the patterns of everyday life and thought—streamlining work routines, editing sequences of real time on film into narratives. The notion of "efficiency"—the final justification industrial society could offer for the changes it brought about—could not have insinuated itself into virtually every walk of life without standardized time. By 1915, and the debate over daylight saving, even leisure and play had to pass the test of efficiency in standard time.

But as Bryan and Darrow's clash at Dayton makes clear, not all Americans accepted the authority of standardized time. A substantial number of citizens—perhaps even a majority in 1925—insisted that time, and the priorities for governing and organizing life and thought, should come from nature. At Dayton, the culture of machines and efficiency ran up against stubborn, entrenched convictions about time and natural law. These were two world views, two simply irreconcilable ways of ordering and making sense of reality.

Are they reconciled today? The answer is not quite clear. Certainly nearly everyone accepts standard time, mandated firmly by law since 1966.[1] But two states—Arizona and Indiana—still refuse to adopt daylight saving, and Congress still gets letters from citizens who regard changing the clock twice a year as an abomination. Even if the issue has been largely settled in the United States, it remains a powerful force in international politics. Some religions—Orthodox Judaism, for example, and Islam—demand observances based on nature and not mechanical time. Muslims must pray at sunrise, noon, and sunset, not at clock hours; and when Ayatollah Khomeini seized power in Iran, he attacked standard time zones as a Western evil and a threat to the virtue of the Islamic nation.

American culture gives more subtle evidence of mixed feelings about time. Americans buy digital watches, which emphasize a linear time—the numbers tick irrevocably toward the future without showing, as an old-fashioned watch dial does, where they were in the past, a few minutes ago. But Americans also love to buy whimsical, silly fashion watches, like the Swiss-made Swatch, which comes in a profusion of constantly changing styles. One recent Swatch model had a black face, no numbers, and black hands—practically unreadable. But you can't make jokes about something your audience doesn't understand, and the popularity of Swatches and their imitators only confirms the pervasiveness of standard time. On TV, through international computer and phone links, at work and rest, we live surrounded by the network of standardized timekeeping. The old devices, time balls or street clocks, have largely disappeared, or else become charming relics buried in museums.

Yet behavioral scientists have discovered patterns of sleep, of alertness, of fertility that run independently of clocks. The human body itself is a clock, with its own internal tickings and hourly indications, and it responds to cues from the natural world. It demands that we respect our internal natural cycles, the circadian rhythms that govern biology. Of course, the same behavioral scientists are hard at work figuring out ways to overcome this internal time. Experiments aimed at fooling the body into overcoming jet lag by controlled exposure to light recently received national press coverage. This line of work only perpetuates the increasing merger between clocks and watches and their ostensible human masters.[2]

It is possible to imagine an industrial society governed by priorities drawn from natural cycles and biological time. Suppose, for example, that American business practices were restructured to allow men and women to fully share work and child rearing. Truly flexible work hours could make it easier to share babysitting—a child's father staying home till three, then working till eleven, for example, while the mother worked during the day. Computers, fax machines, and telephones could make office work at home a realistic possibility. This might not be a better society, or a more pleasant one, but it would be a society that based its decisions about time use on "natural," biological imperatives, not the schedules of the clock.

The history of American time shows that like other values we tend to take as eternal certainties, time is for the most part a plastic, changeable notion, a social creation. Our ideas about what time is have changed over history, and will keep changing as we confront the mixed legacies of industrialization. It seems very likely, for example, that as global concern for the environment grows, we will have to develop renewed respect for the rhythms and cycles of natural time. Will we move away from clocks and watches and back toward nature's example? Only time will tell.

Notes

Abbreviations

ACR *Appendix to the Congressional Record*

AHR *American Historical Review*

AP Papers of William F. Allen, Manuscript Division, New York Public Library

BAC Business Americana Collection, Archives Center, Smithsonian Institution National Museum of American History, Washington, D.C.

CP Papers of George Creel, Manuscript Division, Library of Congress, Washington, D.C.

CPI Records of the Committee on Public Information, 1917–1919, National Archives, Washington, D.C.

CR *The Congressional Record*

DEI Files of the Section of Mechanisms of the Division of Engineering and Industry, Smithsonian Institution National Museum of American History, Washington, D.C.

DP Papers of Charles F. Dowd, Manuscript Division, Boston Public Library

GP Set of microfilm reels, "Frank B. Gilbreth, Selected Papers, ca. 1910–1924," from the Gilbreth Library, Purdue University Libraries, Lafayette, Indiana

LP Papers of Samuel P. Langley, Smithsonian Institution Archives Center, Arts and Industries Building, Washington, D.C.

NA National Archives, Washington, D.C.

Preface

1. *Time*, vol. 133, no. 17 (April 24, 1989), pp. 58–67.

2. On time and clocks in medieval Europe, see David Landes, *Revolution in Time: Clocks and the Making of the Modern World* (Cambridge, MA, 1983); Jacques Le Goff, *Time, Work and Culture in the Middle Ages* (Chicago, 1980), pp. 29–34; Carlo Cippolla, *Clocks and Culture, 1300–1700* (New York, 1977); Lynn White, *Medieval Religion and Technology* (Berkeley, CA, 1978) and *Medieval Technology and Social Change* (New York, 1962). Otto Mayr, *Authority, Liberty and Automatic Machinery in Early Modern Europe* (Baltimore, 1986), discusses the relationship between time, clocks, politics, and machine design in England and Europe.

3. Max Weber, *The Protestant Ethic and the Spirit of Capitalism* (New York, 1976). Anthony Giddens's introduction to this edition offers a review of the various critiques of Weber's thesis. See also Le Goff, *Time, Work and Culture in the Middle Ages*, pp. 29–34, for an example of changing time sense among Catholic merchants, and Edmund Morgan, *American Slavery, American Freedom* (New York, 1975), chs. 3–6, and "The Labor Problem at Jamestown, 1607–1608," in *American Historical Review*, LXXVI (June 1971), pp. 595–611, for examples of "unethical" Protestants.

4. See Michael O'Malley, "The Idea of Time in American Culture" (Ph.D. diss., Berkeley, 1988), introduction. On time and politics, see John G. Gunnell, *Political Philosophy and Time: Plato and the Origins of Political Vision* (reprinted ed. Chicago, 1987); J. G. A. Pocock, *Politics, Language and Time: Essays on Political Thought and History* (New York, 1971); and Hannah Arendt, "Tradition and the Modern Age" and "The Concept of History: Ancient and Modern," in *Between Past and Future* (reprint ed. New York, 1985). On time and literature, see Paul Ricoeur, *Time and Narrative*, vols. I–III (Chicago, 1988), esp. vol III, pp. 3–22; Samuel L. Macey, *Clocks and the Cosmos: Time*

in Western Life and Thought (Hamden, CT, 1980); and Georges Poulet, *Studies in Human Time* (Baltimore, 1956). Useful sociological studies include Michael Young, *The Metronomic Society: Natural Rhythms and Human Timetables* (Cambridge, MA, 1988); Pitirim A. Sorokin and Robert K. Merton, "Social Time: A Methodological and Functional Analysis," in *American Journal of Sociology,* XLII (March 1937), pp. 615–629; Eviatar Zerubavel, *The Seven Day Circle: History and the Meaning of the Week* (New York, 1985) and *Hidden Rhythms: Schedules and Calendars in Social Life* (Berkeley, CA, 1985). On science and time, see Richard Morris, *Time's Arrows: Scientific Attitudes Toward Time* (New York, 1985); J. T. Fraser, *Time, the Familiar Stranger* (Amherst, MA, 1987), pp. 222–288; and Fraser, ed., *The Voices of Time: A Cooperative Survey of Man's Views About Time as Expressed by the Sciences and the Humanities* (Amherst, MA, 1981).

5. Historians are currently debating about the "new social history," which brought the lives of women and ethnic and cultural minorities into the historical mainstream for the first time. Some defend this as healthy cultural pluralism, enriching our knowledge of the past, while others attack it for supposedly fragmenting the practice of history and making it unintelligible. See *American Historical Review,* 94 (June 1989). Understanding and even sympathizing with the desire for consensus, I wish to declare myself unequivocally on the side of the "new" history. This book discusses nearly universal changes, but it doesn't claim that all experienced them in the same way. It traces abstract, standardized public time back to what I believe to be its American origins, but makes no claims that the ideas and beliefs that led to standard time were universally held. At best, this book might help future studies use attitudes about time to explore the complexity and diversity of American culture.

I

Time, Nature, and the Good Citizen

1. New Haven *Columbian Weekly Register,* Oct. 21, 1826, p. 3. Same anonymous letter reprinted in *Connecticut Herald,* Oct. 24, 1826, p. 3.

2. Timothy Dwight, *Travels in New England and New York, 1821–1822* (reprint ed. Cambridge, MA, 1969), vol. 1, pp. 139, 141.

3. Genesis 1:5, 14: "Let there be lights in the firmament of the heavens . . . and let them be for signs, and for seasons, and for the days and years."

4. *Columbian Weekly Register,* Oct. 21, 1826, p. 3.

5. *Connecticut Journal,* Nov. 14, 1826.

6. Quoted in Milton Drake, *Almanacs of the United States,* 2 vols. (New York, 1962), vol. 1, p. ix. For Eli Terry, see Chauncey Jerome, *History of the American Clock Business for the Past Sixty Years and Life of Chauncey Jerome* (New Haven, 1860), pp. 5–17; Penrose R. Hoopes, *Connecticut Clockmakers of the Eighteenth Century* (New York, 1930), p. 113; Carl Dreppard, *American Clocks and Clockmakers* (New York, 1947), ch. VI; and Chris Bailey, *Two Hundred Years of American Clocks and Watches* (Englewood Cliffs, NJ, 1975).

7. *Connecticut Journal,* Dec. 5, 1826.

8. Ibid., Dec. 12, 1826.

9. Ibid., Feb. 6, 1827, p. 2.

10. Ibid., Mar. 6, 1827, p. 2. On Terry, see Hoopes, *Connecticut Clockmakers,* p. 113.

11. *Connecticut Journal,* Mar. 27, 1827. Other New Haven papers are silent on the matter as well. After 1830, perhaps influenced by the controversy, Terry began pasting equation tables inside his ordinary mean time clocks, so that buyers could make their own adjustments as they saw fit.

12. Relevant works in the vast literature include Mary P. Ryan, *Cradle of the Middle Class: The Family in Oneida County, New York, 1790–1865* (New York, 1984); Michael Rogin, *Subversive Genealogy: The Politics and Art of Herman Melville* (Berkeley, CA, 1983); David Rothman, *The Discovery of the Asylum* (Boston, 1971); Paul Johnson, *A Shopkeepers Millennium* (New York, 1978); Jonathan Prude, *The Coming of Industrial Order* (Cambridge, 1983); Sean Wilentz, *Chants Democratic* (New York, 1984); and Alan Dawley, *Class and Community* (Cambridge, MA, 1982).

13. On the early history of clocks and timekeeping, see Carlo Cipolla, *Clocks and Culture, 1300–1700* (New York, 1977), pt. 1, and David

Landes's remarkable *Revolution in Time: Clocks and the Making of the Modern World* (Cambridge, MA, 1983), pts. 1 and 2.

14. Hoopes, *Connecticut Clockmakers*, gives some description of early American timekeeping, as does Dreppard, *American Clocks and Clockmakers*, and Bailey, *Two Hundred Years of American Clocks*.

15. Quoted in Laurel Thatcher Ulrich, *Good Wives: Image and Reality in the Lives of Women in Northern New England, 1650–1750* (New York, 1983), p. 78. For a more detailed critique of "task orientation" and its uses in history, see Michael O'Malley, "The Idea of Time in American Culture" (Ph.D. diss., Berkeley, 1988), pp. 5–12. Most studies of time in preindustrial cultures mistake the absence of clocks for indifference to time, and fail to see the coercive or compelling social and cultural forces that govern work. See Peter Rigby, "History and Time," in *Persistent Pastoralists: Nomadic Societies in Transition* (London, 1985), pp. 67–91. The most famous use of "task orientation" is probably E. P. Thompson's influential "Time, Work-Discipline, and Industrial Capitalism," in *Past and Present*, 38 (December 1967), pp. 56–97. Thompson errs by not paying enough attention to the strictures that governed time use before the factory system, and by assuming that all preindustrial societies experience the transition from "task orientation" to industrial time in the same way. See especially Thomas Smith, "Peasant Time and Factory Time in Japan," *Past and Present*, 111 (May 1986), pp. 165–197.

16. T. H. Breen's discussion of Washington the farmer in *Tobacco Culture* (Princeton, NJ, 1985), p. 55. See also Breen, *Puritans and Adventurers* (New York, 1980), pp. 164–196. On the applicability of the Protestant Ethic, see Edmund Morgan, *American Slavery, American Freedom* (New York, 1975), chs. 3–6, and "The Labor Problem at Jamestown, 1607–1608" in *American Historical Review*, LXXVI (June 1971), pp. 595–611.

17. A clear and typically concise description of the duality of Western time appears in Stephen Jay Gould, *Time's Arrow, Time's Cycle: Myth and Metaphor in the Discovery of Geological Time* (New York, 1986), pp. 10–16. Thomas Cole's four-painting sequence *The Voyage of Life* represents the coexistence of cyclical and linear time as it was com-

monly understood in antebellum America. See Barbara Novak's discussion of the paintings in *Nature and Culture: American Landscape Painting, 1825–1875* (New York, 1980), pp. 109–110.

18. On American almanacs, see Herbert Leventhal, *In the Shadow of the Enlightenment: Occultism and Renaissance Science in Eighteenth Century America* (New York, 1976), and David D. Hall, *Worlds of Wonder, Days of Judgment: Popular Religious Belief in Early New England* (New York, 1989), pp. 58–61. See also Marion B. Stowell, *Early American Almanacs: The Colonial Weekday Bible* (New York, 1977); George L. Kittredge, *The Old Farmer and His Almanac* (Boston, 1904); Robb Sagendorph, *America and Her Almanacs: Wit, Wisdom, and Weather, 1639–1970* (Boston, 1970); and Robert T. Sidwell, "The Colonial American Almanacs: A Study in Non-Institutional Education" (Ph.D. diss., Rutgers, 1965). See also Jon Butler, "Magic, Astrology, and the Early American Religious Heritage," in *AHR*, 84 (April 1979), pp. 329–330.

19. Quoted in Stowell, *Early American Almanacs*, p. xiv.

20. Ibid., p. 180. On the role of science and Newtonian mechanics, see Brook Hindle, *The Pursuit of Science in Revolutionary America* (Chapel Hill, NC, 1956); Thomas S. Kuhn, *The Copernican Revolution* (New York, 1959); and Herbert M. Morais, *Deism in Eighteenth Century America* (New York, 1960).

21. Mather described in Leventhal, *Shadow of the Enlightenment*, p. 41. The drawing helped in treating farm animals as well. It divided the body into twelve regions—heart, belly, reins [ribs], secrets [genitals], thighs, knees, legs, feet, head, neck, arms, and breast. While the "anatomy" was often attacked as foolish superstition, it was just as often defended. Some claimed that an almanac without the "man of signs" was practically unsalable. See Butler, "Magic, Astrology and the Early American Religious Heritage," pp. 330–332. Modern collections of folklore and superstition, as Leventhal noted, abound with similar interpretations of natural portents.

22. Kittredge, *The Old Farmer and His Almanac*, p. 81.

23. *Old Farmer's Almanac*, April 1832.

24. William Holmes McGuffey, *The Eclectic Second Reader* (Cincinnati, 1836), pp. 16–17.

25. Ruth Miller Elson, *Guardians of Tradition: American Schoolbooks of the Nineteenth Century* (Lincoln, NE, 1964), p. 19.

26. *McGuffey's Third Reader* (Cincinnati & New York, 1879) pp. 56–57. This particular lesson appears in an 1879 *Reader* and not in an 1853 edition. It may have appeared in an earlier edition I could not locate, although having read a great number of antebellum schoolbooks I regard it as typical of the earlier period.

27. Lyman Cobb, *Cobb's Sequel to the Juvenile Readers* (New York, 1834), p. 16.

28. Lindley Murray, *Sequel to the English Reader* (4th ed., Philadelphia, 1810).

29. John E. Lovell, *Lovell's Progressive Readers* (New Haven, 1857), no. 3, pp. 28–30.

30. Elson, *Guardians of Tradition*, p. 19.

31. Noah Webster, *The American Spelling Book* (New Haven, CT, 1831), p. 57.

32. Information on clocks and clockmakers is dominated by the work of collectors. See below and Dreppard, *American Clocks and Clockmakers*, pp. 29–38, ch. 11; Hoopes, *Connecticut Clockmakers*; and Bailey, *Two Hundred Years of American Clocks*.

33. Brook Hindle and Stephen Lubar, *Engines of Change: The American Industrial Revolution* (Washington, D.C., 1986), p. 220.

34. On the time of city life in early modern Europe, see Jacques Le Goff, *Time, Work and Culture in the Middle Ages* (Chicago, 1980), pt. 1; Carlo Cipolla, *Clocks and Culture, 1300–1700* (New York, 1977), pt. 1; and David Landes, *Revolution in Time* (Cambridge, MA, 1983), pts. 1 and 2.

35. Arthur H. Cole, "The Tempo of Mercantile Life in Colonial America," *Business History Review*, 33 (Autumn 1959), pp. 277–299. On theater and public festival, see Lawrence Levine, *Highbrow/Lowbrow. The Emergence of Cultural Hierarchy in America* (New York, 1988). On labor, see Herbert Gutman, *Work, Culture and Society in Industrializing America* (New York, 1966), ch. 1.

36. Hawthorne's musings have of course been admirably described in Leo Marx, *The Machine in the Garden: Technology and the Pastoral Ideal in America* (New York, 1964). For other examples of the village

clock's connection to virtue and natural harmony, see Hawthorne, "Sundays at Home" and "Sights from a Steeple," in *Tales and Sketches*, Roy Harvey Pearce, ed. (New York, 1982). On early public clocks, see Hoopes, *Connecticut Clockmakers*, pp. 37–41.

37. Query VI of Jefferson's *Notes on the State of Virginia*, quoted in George H. Eckhardt, *Pennsylvania Clocks and Clockmakers* (New York, 1955), p. 35.

38. Quote comparing the Constitution to a clock from Michael F. Lienesch, "The Concept of Time in American Political Thought" (Ph.D. diss., UC Berkeley, 1977), p. 303. On the Constitution, see Michael Kammen, *A Machine That Would Go of Itself* (New York, 1986), and Marx, *The Machine in the Garden*, p. 165. For the connection of time and the clock to European political theory and science, see Otto Mayr, *Authority, Liberty, and Automatic Machinery in Early Modern Europe* (Baltimore, 1986).

39. Jefferson clearly believed in progress, but he just as clearly expected an inevitable decadence. The emphasis on Rome envisions historical cycles on a grand scale, and indeed he and many of his peers interpreted history as a record of rise, decline, and fall. He never believed that America's decline could be permanently avoided; he only hoped to postpone it. See Drew McCoy, *The Elusive Republic: Political Economy in Jeffersonian America* (New York, 1980), esp. ch. 5; Lienesch, "Concept of Time in American Political Thought"; and Major L. Wilson, "The Concept of Time and the Political Dialogue in the United States," in *American Quarterly*, XIX (1967), pp. 619–644. Wilson follows the tension between cyclical and linear time into the Jacksonian period.

40. McCoy, *The Elusive Republic*, p. 136.

41. Quote from Marx, *Machine in the Garden*, p. 161.

42. Henry F. May, *The Enlightenment in America* (New York, 1978), p. 24.

43. See Jerome, *History of the American Clock Business*, p. 16.

44. B. A. Botkin, *A Treasury of American Folklore* (New York, 1947), says that "the yarns about Yankee clock peddlers are legion" (p. 29). For an example, see Thomas Chandler Haliburton, *The Clockmaker* (New York, 1871), and John Cawelti's description of Haliburton's

and other characters in *Apostles of the Self-Made Man* (Chicago, 1965), pp. 69–70.

45. The only scholarly study of Terry's career is John Joseph Murphy's article, "Entrepreneurship in the Establishment of the American Clock Industry," *Journal of Economic History*, XXVI (June 1966), pp. 169–186, and his Ph.D. dissertation, "The Establishment of the American Clock Industry" (New Haven, 1961).

46. George Featherstonhaugh, quoted in Richard D. Brown, "Modernization and the Modern Personality," in *Journal of Interdisciplinary History*, 2 (1972–73), p. 220.

47. Jerome, *History of the American Clock Business*, p. 23.

48. An example appears in Alan Dawley, *Class and Community: The Industrial Revolution in Lynn* (Cambridge, MA, 1982). The illustration on p. 43 depicts a shoemaker's "ten footer" with a typical forerunner of Terry's "pillar and scroll" design.

49. Newman Ivey White, ed., *The Frank Brown Collection of North Carolina Folklore* (Durham, NC, 1961), no. 5049; Wayland Hand, et. al., eds., *Popular Beliefs and Superstitions: A Compendium of American Folklore* (Boston, 1981), nos. 27350–27352; Harry M. Hyatt, ed., *Folklore from Adams County, Illinois* (Chicago, 1960), nos. 513–514.

50. White, *Frank Brown Collection*, nos. 5052–5055; Hand, *Popular Beliefs and Superstitions*, nos. 27326–27329, 27337; Hyatt, *Adams County Folklore*, nos. 286–287; Anthon S. Cannon, *Popular Beliefs and Superstitions from Utah* (Salt Lake City, 1984), pp. 291–292; William I. Koch, *Folklore from Kansas* (Lawrence, KA, 1980), no. 1842.

51. Hyatt, *Adams County*, no. 11425; White, *Frank Brown Collection*, no. 3053.

52. Hand, *Popular Beliefs*, nos. 17353–17356; White, *Frank Brown Collection*, no. 3056; Hyatt, *Adams County*, nos. 14950, 11433; Cannon, *Popular Beliefs and Superstitions from Utah*, no. 6468.

53. Henry Wadsworth Longfellow, *Evangeline and Selected Tales and Poems*, Horace Gregory, ed. (New York, 1964), pp. 42–43.

54. *McGuffey's Third Reader*, pp. 137–138. See also "The Old House Clock" in *McGuffey's Fifth Reader*, pp. 232–233; "The Kitchen Clock" in *McGuffey's Second Reader*, pp. 78–81.

55. Jesse Olney, *The National Preceptor* (New York, 1839), p. 221.

56. Mark Twain, *The Adventures of Huckleberry Finn* (1884) in Guy Cardwell, ed., *Mississippi Writings* (New York, 1982), pp. 724–727.

57. Thanks to Michael Rogin for help with this material. On sentimental culture, see Anne Douglas, *The Feminization of American Culture* (New York, 1977), esp. pp. 17–139, 241–272; Karen Halttunen, *Confidence Men and Painted Women: A Study of Middle-Class Culture in America, 1830–1870* (New Haven, 1982). On the decline of paternalism, see Mary P. Ryan, *Cradle of the Middle Class: The Family in Oneida County, New York, 1790–1865* (New York, 1984), chs. 4 and 5; David Rothman, *The Discovery of the Asylum* (Boston, 1971); and Paul Johnson, *A Shopkeepers Millennium* (New York, 1978).

58. John Pierpont, *American First Class Book* (Boston, 1831), pp. 314–316. See also *McGuffey's Fifth Reader*, pp. 114–116. Like a lot of material in readers, the story came from England. The two stories reverse our common understanding of women's relationship to nature. In the former, men stand for preindustrial patriarchal authority and the cyclical, natural inevitability of birth, youth, aging, and death. In the latter, women, as "Mistress Dial," represent the regulating industrial machine and the virtues required for progress. See discussion of Catherine Beecher below.

59. Herbert Gutman, *Work, Culture and Society in Industrializing America* (New York, 1966), p. 26.

60. Anthony F. C. Wallace, *Rockdale: The Growth of an American Village in the Early Industrial Revolution* (New York, 1980), pp. 328, 179. I am not sure if or how cotton mills were lighted after dark, except that Lowell was lighted by arc lamps in the 1830s. The hours listed in mill timetables are deceptive—it is now difficult for us to guess how much daylight was left, for example, in Pawtucket, RI, at 7:30 p.m. in July, thanks to time zones and daylight saving time.

61. Gutman, *Work, Culture and Society*, p. 27. Gutman points out that collective protest at Lowell began only after a "rationalization" in the 1840s, when stricter time schedules appeared.

62. E. P. Thompson, "Time, Work-Discipline, and Industrial Capitalism," in *Past and Present*, 38 (December 1967), pp. 56–97. Among the many historical studies building on Thompson's anaylsis are Herbert Gutman's enormously influential *Work, Culture and Society* *in Industrializing America* (New York, 1966), ch. 1; Jonathan Prude, *The Coming of Industrial Order* (Cambridge, MA, 1983), pp. 15–17, 36–37, 130–131; and Alan Dawley, *Class and Community* (Cambridge, MA, 1982).

63. "Report of the Committee on Education of the New England Association of Farmers, Mechanics and Other Workingmen," printed in *The Free Enquirer*, June 1832, and reprinted in Gary Kulik, ed., *The New England Mill Village* (Cambridge, MA, 1982), pp. 495–496. See also Kulik, "Pawtucket Village and the Strike of 1824: The Origins of Class Conflict in Rhode Island," in *Radical History Review*, 7 (Spring 1978), p. 28.

64. Pottsville, Pennsylvania, *Miner's Journal and Pottsville Daily Advertiser*, Oct. 21, 1843, p. 2; Dec. 16, 1843, p. 2.

65. "Witnesses in the Contested Election of Jeremiah Reed," Dec. 8, 1843, in Christopher Loeser Papers, Schuylkull County Historical Society, Pottsville, PA, v. 24, pp. 73–74. I am greatly indebted to Grace Palladino of the Samuel Gompers Papers for sharing this information with me, and to John Joy of the Schuylkill County Historical Society for confirming the dates.

66. Kulik, ed., *The New England Mill Village*, pp. 265–266.

67. Edward Everett, *An Address Delivered Before the Massachusetts Charitable Mechanics Association, 20th September 1837, On the Occasion of Their First Exhibition and Fair* (Boston, 1837), p. 18. Thanks to Carlene Stephens for bringing the address to my attention.

68. Lydia Maria Child, *The American Frugal Housewife* (Boston, 1835), p. 1.

69. William Alcott, *The Young Housekeeper* (Boston, 1838), p. 61.

70. Ibid., p. 63.

71. Ibid., pp. 64–65.

72. Nathaniel Hawthorne, "Time's Portraiture," in *Tales and Sketches*, Roy Harvey Pearce, ed. (New York, 1982), p. 587. The Elgin Watch Co., after the Civil War, adopted as its symbol Father Time discarding his hourglass for a pocketwatch, as in Hawthorne's story.

73. "The Artist of the Beautiful," in *Tales and Sketches*, Pearce, ed., pp. 910, 914.

74. Edgar Allan Poe, *Poetry and Tales,* Patrick F. Quinn, ed. (New York, 1984), pp. 298–306. Poe satirized the navigator's confusion over time standards in "Three Sundays in a Week" (1841).

75. It was during these years that it first became possible to electrically synchronize one clock, later called the "master" clock, with another, less accurate "slave" clock. See Chapter Two.

76. Poe, *Poetry and Tales,* pp. 491–505. In "The Pit," the enemy is the Inquisition, the fusion of Church and State power run amok. It suggests Poe's hostility to the secularization of time, or perhaps the perversion of Time, an abstract, religious quality, by secular power. Poe used the motif of the clock more than once. "A Predicament," part 2 of the 1838 parody "How to Write a Blackwood Article," describes a woman decapitated by the razor-sharp hands of an enormous tower clock. In "The Colloquy of Monos and Una," 1841, Poe insisted on a higher concept of time and chronology, perceptible only to the soul and perverted by crude mechanical timekeepers.

77. Robert T. Handy, *A Christian America: Protestant Hopes and Historical Realities* (New York, 1985), p. 45. See also George Marsden, *Fundamentalism and American Culture* (New York, 1980), p. 13.

78. Daniel Wilson, *The Divine Authority and Perpetual Obligation of the Lord's Day Asserted in Seven Sermons* (New York, 1831), p. 173. See also John Holmes Agnew, *A Manual of the Christian Sabbath* (Philadelphia, 1834); John S. Stone, *Lectures on the Institution of the Sabbath* (New York, 1844); J. B. Waterbury, *A Book for the Sabbath* (New York, 1840); and Silas M. Andrews, *The Sabbath at Home* (Philadelphia, 1836).

79. Handy, *A Christian America,* p. 232, n. 58: "Sabbatarians usually argued that without Sabbath rest prosperity would fail." An example of hostility to the railroads, in the form of a petition to the Massachusetts legislature asking for a ban on Sunday travel, appears in *Proceedings of the Anti-Sabbath Convention* (Boston, 1848; reprint ed. (New York, 1971), pp. 116–117.

80. *Proceedings of the Anti-Sabbath Convention,* pp. 39–40.

81. Ibid., pp. 113–114.

82. Catherine Beecher, *A Treatise on Domestic Economy* (1841), (New York, 1970), pp. 170–175.

83. On Beecher's background and motives, see Katherine Kish Sklar, *Catherine Beecher: A Study in American Domesticity* (New Haven, 1973).

84. Beecher, *Treatise on Domestic Economy*, p. 109. A diligent search disclosed only one sentence in any edition mentioning clocks.

85. This is not to claim Beecher originated the split between home time and work time, only that she was its most articulate spokesperson.

86. Henry A. Miles, *Lowell As It Was, and As it Is* (Boston, 1845), pp. 65–66, quoted in John F. Kasson, *Civilizing the Machine: Technology and Republican Values in America, 1776–1900* (New York, 1976), pp. 75–76. Kasson points out the supervisor's strategic location between the dormitories and the mills. The clock tower, on the other hand, overlooked all, and the association with the city of Lowell made factory time and city time the same.

87. Kasson, *Civilizing the Machine*, pp. 103–106.

II
Celestial Railroad Time

1. Determining longitude involves comparing local time, measured astronomically, with the time at another location. Comparing the two times gives the two points' distance apart. At first highly accurate, portable clocks or chronometers were used, but the telegraph made them less necessary. On the relationship between longitude, navigation and timekeeping, see Derek Howse, *Greenwich Time and the Discovery of the Longitude* (Oxford, 1980).

2. On Mason and Dixon, see Brook Hindle, *The Pursuit of Science in Revolutionary America* (Chapel Hill, NC, 1956), pp. 166–167. On Lewis and Clark, see William H. Goetzmann, *New Lands, New Men: America and the Second Great Age of Discovery* (New York, 1986), pp. 110–115, and John C. Greene, *American Science in the Age of Jefferson* (Ames, IA, 1984), pp. 196–197. Not all surveying demanded clocks, but the most accurate measurements did.

3. Greenwich Observatory was closely associated with accurate time and navigation, having originally been established to aid English navigators. As early as 1714, Parliament had offered a prize for a clock capable of keeping accurate Greenwich time at sea. See Howse, *Greenwich Time.*

4. J. E. Nourse, "Memoir of the Founding and Progress of the United States Naval Observatory," in I. Bernard Cohen, ed., *Aspects of Astronomy in America in the Nineteenth Century* (New York, 1980), p. 6. On the Coast Survey, see Robert V. Bruce, *The Launching of Modern American Science* (New York, 1987), pp. 171–186; the *Annual Reports* of the Superintendent of the Coast Survey, 1846–51; and Willis Milham, "Early American Observatories," in Cohen, ed., *Aspects of Astronomy*, pp. 8–10.

5. Nourse, "Memoir of the Founding," in Cohen, ed., *Aspects of Astronomy*, p. 10.

6. Ordinary surveyors did not carry chronometers. But mapping a territory requires some stable reference point from which to refer subsequent measurements. The Coast Survey's work offered such reference points for routine, less precise surveying. On disputes over title and western expansion, see Michael Rogin, *Fathers and Children: Andrew Jackson and the Subjugation of the American Indian* (New York, 1976), ch. 3. On the rationalization of mapping and land use and its relation to culture, see Philip Fischer, "Democratic Social Space: Whitman, Melville and the Promise of American Transparency," in *Representations*, 24 (Fall 1988), pp. 60–101; and John Stillgoe, *Common Landscape in America* (New York, 1977).

7. One historian of science even compared the enthusiasm for astronomy with the twentieth century's passion for Lindbergh. G. Edward Pendray, "How the Telescope Came to America," in Cohen, ed., *Aspects of Astronomy*, pp. 210–212, and Bruce, *Launching of Modern American Science*, passim.

8. John Quincy Adams, *An Oration Delivered Before the Cincinnati Astronomical Society on the Occasion of Laying the Cornerstone of an Astronomical Observatory* (Cincinnati, 1843), pp. 12–18, quote at p. 39.

9. Ibid., p. 68.

10. Adams's metaphor also recalls once again "Mistress Dial" of the children's story "The Discontented Pendulum," and suggests that time is a female principle organizing domestic production, especially the production of good democratic citizens. See Chapter One.

11. Allan R. Pred, *Urban Growth and the Circulation of Information: The United States System of Cities, 1790–1840* (Cambridge, MA, 1973), p. 37.

12. Quoted in Hugo Arthur Meier, "The Technological Concept in American Social History" (Ph.D. diss., Wisconsin, 1951), p. 85.

13. Pred, *Urban Growth and the Circulation of Information*, pp. 45, 53, 176–177.

14. Quoted in Carlene Stephens, " 'The Most Reliable Time': William Bond, the New England Railroads, and Time Awareness in 19th Century America," in *Technology and Culture*, 30 (January 1989), p. 5. Stephens points out that the postal service was far more interested in punctuality—and more punctual—than the early railroads. Thanks also to Coleen Dunlavy of the University of Wisconsin at Madison and Steven Lubar of the National Museum of American History for information on early railroads.

15. Allan R. Pred, *Urban Growth and City-Systems in the United States, 1840–1860* (Cambridge, MA, 1980), p. 149.

16. Nathaniel Hawthorne, *Tales and Sketches*, Roy Harvey Pearce, ed. (New York, 1982), pp. 808–824. The short story form, compacting Bunyan's lengthy book into a few pages, sharpens Hawthorne's pointed critique of industrial society. The railroad and its quickening pace have even changed the terms of communication—the railroad travels Bunyan's narrative landscape in fraction of the time.

17. Edward Everett, "The Uses of Astronomy," in Cohen, ed., *Aspects of Astronomy*, p. 19.

18. Since the railroad was significantly faster than the apparent course of the sun, and every fifteen degrees of longitude equals one hour of time, rapid east-to-west travel made regional time differences more glaring. See Elias Loomis, *The Recent Progress of Astronomy*, 3d. ed. (New York, 1856), p. 288.

19. For example, see *Appleton's Railroad and Steam Companion*, first published in 1848, and *Dinsmore's Railroad and Steam Navigation Guide*, 1853–1863.

20. Everett, "Uses of Astronomy," p. 19. See also Loomis, *Recent Progress of Astronomy*, pp. 287–292.

21. Stephens, " 'The Most Reliable Time,' " pp. 1–24.

22. Quoted in Michael P. Rogin, *Subversive Genealogy: The Politics and Art of Herman Melville,* (Berkeley, CA, 1985), p. 138.

23. Stephens, " 'The Most Reliable Time,' " p. 7. There are several conflicting contemporary accounts of exactly what time the railroads adopted. The best history of the Bonds' work is Stephens's article and her monograph in preparation on William Bond and Son. See also Bessie Zaban Jones and Lyle Gifford Boyd, *The Harvard College Observatory* (Cambridge, MA, 1971), ch. 2.

24. Leo Marx, *The Machine in the Garden* (New York, 1964), p. 248. Marx explores Thoreau's sophisticated and ambivalent feelings about the railroad in ch. 5, pt. 2, pointing out that Thoreau's feelings were by no means all negative. Background on time and the clock in the writing of the American Renaissance appears in Samuel Macey, *Clocks and the Cosmos: Time in Western Life and Thought* (Hamden, CT, 1980), and in Georges Poulet, *Studies in Human Time* (Baltimore, 1956), pp. 323–349.

25. Brooks Atkinson, ed., *Walden and Other Writings of Henry David Thoreau* (New York, 1950), pp. 106 and 107.

26. Ibid., pp. 111 and 112.

27. Ibid., pp. 105, 107, respectively.

28. *Running Regulations of the Camden and Amboy Railroad,* 1853, on exhibit at the National Museum of American History, Washington, D.C., and reprinted in John H. White Jr., *The John Bull* (Washington, D.C., 1981), p. 133.

29. Such rules also obviously heighten the difference between work time and leisure time—work time means carrying a watch and watching (in both senses of the word) the clock.

30. Herman Melville, *Pierre, or The Ambiguities,* Harrison Hayford, ed. (New York, 1984), pp. 247–252. Melville may also have been related to J. Melville Gillis, in charge of the National Observatory in the 1840s.

31. Howse, *Greenwich Time,* pp. 49–53. See also Nigel Thrift, "Owner's Time and Own Time: The Making of a Capitalist Time Consciousness, 1300–1880," in *Space and Time in Geography,* Allan Pred, ed. (Lund, 1981), pp. 56–84.

32. Nautical almanacs charted the appearances and movements of stars and planets, listing the times of their appearance. The English *Nautical Almanac,* used a reference guide to the heavens by sailors, astronomers, and surveyors since the early nineteenth century, listed its calculations in Greenwich time.

33. William Bond, then engaged in determining the location of Cambridge, Massachusetts, relative to Greenwich, personally collected signatures on a petition demanding the retention of Greenwich time. Though Congress eventually settled on a compromise, an *American Almanac* including calculations in both Greenwich and Washington, D.C., time, the American Prime Meridian was never widely adopted. See Charles Henry Davis, *Remarks Upon the Establishment of an American Prime Meridian* (Cambridge, 1849); Stephens, " 'The Most Reliable Time' "; Craig Waff, "Charles Henry Davis, the Foundation of the American Nautical Almanac, and the Establishment of an American Prime Meridian," in *Vistas in Astronomy,* 28 (1985), pp. 61–66; and *The Congressional Globe,* June 16, 1846, pp. 961–963; May 2, 1850, pp. 891–892. A number of maps were produced showing the meridian of Washington and the Observatory. Several remain in the Library of Congress.

34. Melville, *Pierre,* p. 249.

35. The railroads accepted Bond's standard in lieu of the proposed American Prime Meridian described above. The time was calculated on a Greenwich meridian crossing the middle of Massachusetts Bay and reflected the railroads' solid opposition to the Washington standard. See Stephens, " 'The Most Reliable Time.' "

36. For a fuller explanation of the passage's importance for the novel, see Michael Rogin, *Subversive Genealogy: The Politics and Art of Herman Melville* (Berkeley, CA, 1983), pp. 177–178. "Bartleby the Scrivener" echoes the theme of time and moral coercion. The lawyer's two original copyists are difficult but not unmanageable, because each is tied to "natural" time. Melville took great pains to establish that each one's peronality changed at "twelve o'clock, meridian," or noon by the sun. Bartleby, on the other hand, has no past, no future, no natural appetites, and no schedule—he is nearly "timeless," unregulated and ungovernable. Bartleby's refusal to keep regular hours,

to accept standard customs about the organization of time and self-government, thrusts the bourgeois narrator into an unresolvable moral quandary. In "The Bell Tower," one of the "Piazza Tales" written after the failure of *Pierre*, Melville's protagonist, a driven, monomaniacal artist, constructs a mechanical timekeeper in the shape of a man. The whole town gathers to hear the clock strike, but the clock/man kills the artist in the act of striking, making the artist—and, by implication, Melville—the victim of mechanical time and the regular universe he constructs.

37. "Railroad Time," in *American Railway Journal*, 25 (Aug. 21, 1852), pp. 529–530, 530.

38. J. N. Bowman, "Driving the Last Spike at Promontory, 1869," in *Utah Historical Quarterly*, 37 (Winter 1969), pp. 89–90. Thanks to Anne Hyde for suggesting the ceremony at Promontory.

39. *Proceedings of the American Metrological Society*, II (New York, 1880), p. 18; *Annual Report of the Astronomer, Dudley Observatory* (Albany, NY, 1878), p. 14; *Annual Report of the Director, Cincinnati Observatory* (Cincinnati, 1869), pp. 35–36; Gustavus A. Weber, *The Naval Observatory: Its History, Activities and Organization* (Baltimore, 1926), p. 34; *Annual Report of the Board of Directors of the Chicago Astronomical Society* (Chicago, 1881), p. 1.

40. Donald Leroy Obendorf, "Samuel P. Langley: Solar Scientist, 1867–1891" (Ph.D. diss., Berkeley, 1969), pp. 1–11. Langley was requested to participate in weather surveys, but felt unable to oblige so early in his tenure.

41. Pennsylvania Railroad Company, *General Order No. 4*, Jan. 26, 1871, in microfilm copies of the Langley Papers at the Smithsonian Institution Archives (hereafter referred to as LP, Reel 2, sec. 2). Thanks to the Coast Survey, which had fixed the precise longitude of major cities, Langley could cable Philadelphia time to the eastern divisions of the Pennsylvania, and later Columbus, Ohio, time to the western division.

42. Obendorf, "Samuel P. Langley," pp. 10–12, Langley to the Allegheny Observatory Committee, May 1872 (LP, R. 1, s. 2).

43. *The Pittsburgh Commercial*, Feb. 12, 1870.

44. Carlene E. Stephens, "The Impact of the Telegraph on Public Time in the United States, 1844–1893," in *Technology and Society*, 8 (March 1989), p. 6.

45. Langley to Allegheny Observatory Committee, May 1872 (LP, R. 1, s. 1), p. 5. A provocative discussion of time, space, and money appears in geographer David Harvey's *Consciousness and the Urban Experience: Studies in the History and Theory of Capitalist Expansion* (Baltimore, 1985), ch. 1.

46. Charles F. Dowd, *System of National Time and Railway Time Gazetteer* (Albany, NY, 1870). Dowd's first plan, proposed in 1869, called for Washington time alone as the standard. The idea of zones occurred to him a year later. Dowd's inspiration, or the sources of his ideas, remain obscure. His personal papers, at the Boston Public Library, consist of little more than published accounts of his work. See Charles North Dowd, *Charles F. Dowd and Standard Time* (New York, 1930).

47. Dowd, *A Paper Read Before the American Metrological Society, Dec. 27, 1883* (Saratoga Springs, NY, 1883). The railroad managers who reviewed the plan pronounced it eminently workable but unnecessary. Most customers, they claimed, traveled only short distances and kept time changes in mind. See *Railroad Gazette*, May 24, 1873, p. 206.

48. Dowd wrote Cassatt in October 1870, briefly explaining his plan and enclosing a testimonial from other railroad managers. See Dowd to Cassatt, Oct. 20, 1870 (LP, R. 1, s. 4). Cassatt wrote Thaw that "if the thing is to be done, as of course it will be in one shape or another, we would prefer to have our Pennsylvania Institution have the credit of it, and I therefore thought it better to put Mr. Dowd off with an evasive answer." See Cassatt to Thaw, Oct. 24, 1870 (LP, R. 2, s. 2). Langley also wrote to Edward Pickering of Harvard University warning him that unless astronomers came up with plans of their own, Dowd's might prevail. Dowd eventually abandoned local time, but by then others, better connected, were pursuing the question, and his natural desire to be paid for his efforts made him unwelcome. Thanks to an energetic promotional campaign on his own behalf, however, when standard time was adopted in 1883,

Dowd received credit for its invention from many sources, including *Harper's Weekly,*) 27 (Dec. 29, 1883), p. 843. A mildly hostile and confusing public debate ensued over standard time's origins. See also Charles North Dowd, *Charles F. Dowd,* and William F. Allen, *Short History of Standard Time and Its Adoption in North America* (New York, 1903).

49. Evidence for this comes from a multitude of sources. Besides the example of Bond and Langley, see *Annual Report of the Astronomer, Dudley Observatory* (Albany, NY, 1878), p. 14; *Report of the Director of Harvard College Observatory,* Nov. 14, 1878, pp. 9–10; Obendorf, "Samuel Langley," pp. 14–15; and William F. Allen, *History of the Adoption of Standard Time. Read Before the American Metrological Society, on December 27, 1883* (New York, 1884), pp. 14–15. The best evidence comes from the papers of William F. Allen at the New York Public Library (see Chapter Three). As editor of the *Traveler's Official Guide,* Allen sent circulars to the managers of every American railroad in 1882, asking what standard of time they used. From the responses he drew up a map showing the fifty-three railroad times he found in use in 1882. The responses and the map clearly show the railroads organized into regional time zones. Allen's map is in the archives of the Division of Engineering and Industry of the National Museum of American History. See also Ian R. Bartky, "The Adoption of Standard Time," in *Technology and Culture,* 30 (January 1989), pp. 25–56, and "The Invention of Standard Time," in *Railroad History,* 148 (Spring 1983), pp. 18–20.

50. Dowd, *Paper Read Before the American Metrological Society,* pp. 4–5.

51. John Rodgers to William F. Allen, Feb. 20, 1882, in William F. Allen Papers, Manuscript Division, New York Public Library (hereafter referred to as AP), v. I, p. 10.

52. Several extremely rare examples of agitation for uniform time appeared in the *Railroad Gazette* in 1870. See April 2, p. 6; April 9, p. 29; April 16, p. 31; May 7, p. 1; and May 31, p. 181. The editors commented favorably on the resumption of public time signals from Dearborn Observatory, and also excerpted one of Langley's descriptions of his service (April 9).

53. Compare Dowd, *Paper Read Before the American Metrological Society,* pp. 4–5, with Allen, *History of the Adoption of Standard Time,* pp. 14–

15. Allen claimed that the number had been reduced to fifty-three by 1882.

54. Langley, "The Electric Time Service," in *Harper's New Monthly Magazine*, LVI (April 1878,) pp. 665–671, quote at p. 666. The *Harper's* article provides a fuller explanation. Specifications for the system are in LP. Despite Langley's talk about accuracy, the system never worked all that well.

55. Thaw to Allegheny Observatory Committee, May 15, 1872, quoted in Obendorf, "Samuel Langley," p. 15; Langley to Cassatt, Feb. 20, 1872 (LP, R. 1, s. 1). Obendorf misread Layng's name, understandably given his very poor handwriting, as "Layny."

56. Pittsburgh, Commission on the City Hall, *The City Hall* (Pittsburgh, 1874), pp. 41, 36; Obendorf, "Samuel Langley," pp. 20–23.

57. See below and Langley, *Report of the Director of the Allegheny Observatory* (Pittsburgh, 1873); *Railroad Gazette*, April 9, 1870, p. 29; *Scientific American* (Jan. 13, 1872), p. 34; *American Journal of Science and Arts* (November 1872), pp. 377–386; *Telegraph Engineer's Journal*, 1 (1872–73), pp. 433–441; *The American Exchange and Review*, XXIV (January 1874), pp. 271–276; and the numerous letters from Langley to the Observatory Committee in LP.

58. Langley, "The Electric Time Service," in *Harper's New Monthly Magazine*, LVI (April 1878), pp. 665–671; quotes at pp. 669, 670, 671, 665. The article was adapted from several previous ones.

59. Everett, "The Uses of Astronomy," in Cohen, ed., *Aspects of Astronomy*, p. 17.

60. Langley's imagery is also William Alcott's recommendation of an alarm clock for the restless sleeper (see Chapter One) writ large.

61. Michel Foucault, *Discipline and Punish: The Birth of the Prison* (New York, 1979), pp. 173, 173–175.

62. *Annual Report of the Director of Harvard College Observatory* (Boston, 1877), pp. 9–10; *Annals of the Astronomical Observatory of Harvard College*, III (Boston, 1877), pp. 11–12.

63. Leonard Waldo, *Standard Public Time* (Cambridge, 1877), p. 1; *Report of the Director of Harvard College Observatory* (Boston, 1878), p. 8; Edward C. Pickering, "Time Service of Harvard College Observ-

atory," in *Popular Science Monthly*, 19 (Feb. 12, 1892), p. 88. In 1878 the time ball dropped on Boston local time, although plans to drop it on Greenwich time were afoot.

64. Gustavus A. Weber, *The Naval Observatory: Its History, Activities and Organization* (Baltimore, 1926), p. 34. The terms of the arrangement remain obscure, as does much of the next sixty years' profitable cooperation between the two institutions. Western Union's archives are now in the Archives Center of the National Museum of American History. There is almost nothing on the time service besides a collection of early twentieth-century newspaper clippings.

65. *New York Tribune*, Feb. 3, 1877, p. 6; Mar. 15, 1877, p. 4; *New York Times*, Sept. 8, 1877, p. 2; *Journal of the Telegraph*, 10 (April 16, 1877), p. 114.

66. Edward S. Holden, "On the Distribution of Standard Time," in *Popular Science Monthly*, II (June 1877), pp. 174–182. Holden denied that the Naval Observatory signals interfered or conflicted with any other observatory's. It is not clear what the Naval Observatory gained from the relationship besides the free transmission of some messages. The Observatory may have hoped, by strengthening its position as the source of time signals, to increase its annual appropriation. The enormous political and economic power of Western Union may have been persuasive as well.

67. Langley to Edward Pickering, April 9, 1877, quoted in Obendorf, "Samuel Langley," p. 25.

68. *Pittsburgh Commercial*, Feb. 17, 1870.

69. It must be pointed out that Observatory staff, with few exceptions, never profited personally from time sales. On the other hand, time services freed astronomers from having to scare up funds and earned money for research. Langley profited from his service, which made over $60,000 during his tenure, in that it freed him to pursue the solar researches that made his reputation. The point is not personal greed—most were hardly venal men—but the use of time as a commodity and the attempt to standardize it. On the relationship between science and business, see Bruce, *The Launching of Modern American Science*, passim.

70. Truman Abbe, *Professor Abbe and the Isobars* (New York, 1955), passim.

71. *Annual Report of the Director of the Cincinnati Observatory* (Cincinnati, June 1870). Abbe directed the Cincinnati Observatory.

72. Abbe, *Professor Abbe*, p. 145. The most complete record of these scientists' work, and of American time reckoning habits before 1883, appears in the *Proceedings of the American Metrological Society*, 1–5 (New York, 1889). The Society should not be confused with the American *Meteorological Society*. For a detailed recounting, see Ian R. Bartky, "The Adoption of Standard Time," in *Technology and Culture*, 30 (January 1989), pp. 25–56.

73. Cleveland Abbe (?), "Standard Time in America," in *Science*, 22, n.s. (Sept. 8, 1905), pp. 315–318; American Metrological Society, *Standard Time Circular Number Two* (Mar. 1, 1881), in *Proceedings of the AMS*, II (New York, 1880), p. 181.

74. *Proceedings of the AMS*, II, pp. 308–311; Abbe to William F. Allen, June 14, 1879, in AP, v. 1, p. 2.

75. *Proceedings of the AMS*, pp. 308, 181. For the Signal Service, uniform Greenwich time facilitated national and international weather bulletins and scientific and military dispatches while preserving the time services established by individual observatories. Their time, determined by the clearinghouse plan, was to be more accurate and hence "better."

76. *New York Times*, Dec. 24, 1881, p. 1; Feb. 20, 1882, p. 4; *New York Herald*, Dec. 24, 1881; Jan. 7, 1882, p. 4.

77. *Congressional Record*, (hereafter *CR*), July 27, 1882, pp. 6578–6579. The Observatory, like Western Union, pronounced itself in opposition to uniform time standards for the general public. Its service was to apply only to shipping and transport.

78. Obendorf, "Samuel Langley," p. 26.

79. *Science*, 22, n.s. (Sept. 8, 1905), p. 317; *CR*, July 27, 1882, p. 6579. *House Report*, no. 681 (Mar. 9, 1882), "Meridian Time and Time Balls on Customs Houses." Its failure is somewhat mysterious. It was killed by the unstated objections of eleven congressmen, and never revived. This is mostly speculation on my part, but it probably had something to do with the railroads, who strongly opposed any government legislation affecting their operations and were by then at work on standard time systems of their own (see Chapter Three).

The Signal Service, through the American Metrological Society, strongly opposed the bill as well. See *Proceedings of the AMS*, II (May–December 1881), p. 310.

80. See *Papers in the Matter of the Controversy Between the United States (Signal Bureau) and the Western Union Telegraph Company* (New York, 1874).

81. *Journal of the Telegraph*, Feb. 16, 1882, p. 52.

82. Pittsburgh, Commission on the City Hall, *The City Hall* (Pittsburgh, 1874), pp. 41, 36.

83. Leonard Waldo, "The Distribution of Time," in *North American Review*, 131 (December 1880), p. 530; Holden, "On the Distribution of Standard Time," p. 175.

84. *New York Tribune*, Feb. 3, 1870, p. 6; Mar. 3, 1877, p. 4. The *Tribune* rightly pondered the ambiguity in Western Union's time service circular. Did it propose Washington time for the nation, or both Washington and local time?

85. Waldo, "Distribution of Time," in *North American Review*, 131 (December 1880), p. 528.

86. Ibid.

87. State of Connecticut, *Twenty-Eighth Annual Railroad Commissioners' Report* (January 1881), p. 57.

88. *Annual Report of the Board of Managers, Yale College Observatory, 1881* (November 1879–May 31 1881), pp. 6–22. Detailed accounts of the Bureau's work appear in the first through fourth *Annual Reports of the Director in Charge of the Horological and Thermometric Bureaus of The Winchester Observatory of Yale College* (1880–84), photocopies on file in the Division of Engineering and Industry, National Museum of American History, Washington, D.C. Thanks to Carlene Stephens for sharing these. The Bureau dealt mostly with other corporations. The newly formed Waltham Watch Comany, for example, sent some of its products to Waldo for testing. Its business declined rapidly after the first year.

89. State of Connecticut, *Journal of the House of Representatives*, January Session, 1881, pp. 60–61. See House bill no. 11. The bill made no provisions for enforcement.

90. *Annual Reports, Yale Observatory, 1881–1882*, Appendix, p. 16; Connecticut, *Journal of the House of Representatives*, March 1, 1882, House Joint Resolution no. 95.

91 State of Connecticut, *Twenty-Eighth Annual Railroad Commissioners' Report* (January 1881), p. 58.

92. Ibid., pp. 57, 59.

93. Marx, *The Machine in the Garden*, p. 194, testifies to the popularity of the phrase.

94. Waldo, "Distribution of Time," p. 529; William F. Allen in *Proceedings of the American Railway Association*, 1 (1893), p. 690.

95. *Annual Reports, Yale Observatory* (1884–1885), p. 8.

96. Ibid., 1886–87, p. 6. The Standard Time Company disappeared from New Haven city directories after 1885. It may have been absorbed into Western Union's time service.

97. Ibid,. p. 6; and "Leonard Waldo in Reply to Certain Questions Addressed to Him by Members of the Corporation at their Meeting April 18th 1884." Thanks to Carlene Stephens for showing me a photocopy on file in the Division of Engineering and Industry, National Museum of American History, Washington, D.C.

III

The Day of Two Noons

1. Hamlin Garland, *Main Traveled Roads* (New York, 1891), pp. 55, 62.

2. *Proceedings of the American Railway Association*, I (1893), Appendix, pp. 681–682; Stewart H. Holbrook, *The Story of American Railroads* (New York, 1947), p. 356; Ian R. Bartky, "The Invention of Railroad Time," in *Railroad History*, 148 (Spring 1983), p. 13. Similar conventions had met before, under less specific names.

3. Above and two letters in the William F. Allen Papers, Manuscript Division, New York Public Library (subsequently referred to as AP). See William F. Allen to Albert Fink, Sept. 5, 1882 (AP, v. 1, p. 44); and Allen to P. D. Cooper, Aug. 30, 1882 (AP, v. 1, p. 43). The papers are organized into letterbook volumes and numbered boxes with subject folders.

4. William F. Allen, "Railway Operating Associations: An Address Delivered Before the Graduate School of Business Administration, Harvard University, Jan. 11, 1909," p. 5. Information on Allen from the *Dictionary of American Biography* and William F. Allen, *History of the Adoption of Standard Time. Read Before the American Metrological Society, on December 27, 1883* (New York, 1884), pp. 10–11.

5. Allen, *History of the Adoption of Standard Time,* pp. 8–9. See Chapter Two for examples of the need for these guides. Other guides include *Dinsmore's Railroad and Steam Navigation Guide* (1853–63); *Appleton's Railroad and Steam Companion* (1848), and *Appleton's Illustrated Railway and Steam Navigation Guide* (1858–77).

6. Charles F. Dowd, *A Paper Read Before the American Metrological Society, December 27, 1883* (Saratoga Springs, NY, 1883), pp. 5–9. Allen, *History of the Adoption of Standard Time,* p. 9. Frederick T. Newberry, letter to *Travelers Official Guide,* April 1882, reprinted in *Proceedings of the American Railway Association,* I (1893), pp. 684–685.

7. See Chapter Two.

8. *Annual Reports of the Smithsonian Institution, 1882* (Washington, D.C., 1883), pp. 321–322. American Society of Civil Engineers, *Standard Time for the United States of America, Canada, and Mexico* (New York, 1882). Fleming suggested either adopting the twenty-four-hour notation now used in the military and in Europe, or labeling AM and PM with numbers and letters, or using letters to designate standard and numbers to designate local time.

9. *Standard Time: Replies to Questions Submitted by the Special Committee of the American Society of Civil Engineers* (Ottawa, 1882), in AP, Bx 2, fl. 2.

10. *Senate Reports,* 4 (July 18, 1882), Report no. 840.

11. *Proceedings of the American Association for the Advancement of Science,* 30 (August 1881), p. 5. Several members of this group later adopted the American Metrological Society's plan.

12. Association for the Reform and Codification of the Law of Nations, Committee on Standard Time, *Views of the American Members of the Committee as to the Resolutions Proposed at Cologne Recommending a Uniform System of Time Regulation for the World* (New York, 1882), pp. 5–7.

13. Abbe to Allen, June 14, 1879 (AP, v. 1, p. 2); *Proceedings of the American Metrological Society*, II, 1878–79 (New York, 1880), p. 29. Allen, *History of the Adoption of Standard Time*, p. 11.

14. Allen, *History of the Adoption of Standard Time*, p. 8. On regional time standards, see Chapter Two and below.

15. On Connecticut, see Chapter Two and *Hartford Courant*, Jan. 25, 1881, p. 2; Feb. 17, p. 2; Feb. 25, p. 1; Mar. 4, p. 1. On the Signal Service, see Chapter Two and *New York Times*, Dec. 24, 1881, p. 1; Feb. 20, 1882, p. 4; *New York Herald*, Dec. 24, 1881; Jan. 7, 1882, p. 4.

16. "General Von Moltke on Time Reform," in Canada, Parliament, *Documents Relating to the Fixing of a Standard of Time and the Legislation Thereof* (Ottawa, 1891), pp. 25–28. See also Howse, *Greenwich Time and the Discovery of the Longitude* (Oxford, 1980), p. 120; and Stephen Kern, *The Culture of Time and Space* (Cambridge, MA, 1983), pp. 12–16.

17. Howse, *Greenwich Time*, p. 131.

18. Malcolm M. Thompson, *The Beginning of the Long Dash: A History of Timekeeping in Canada* (Toronto, 1978), pp. 33–35.

19. Howse, *Greenwich Time*, pp. 134–138.

20. Association for the Reform and Codification of the Law of Nations, *Views of the American Members*, pp. 14–15. International Meridian Conference, *International Conference Held at Washington for the Purpose of Fixing a Prime Meridian and a Universal Day. October, 1884. Protocols of the Proceedings* (Washington, D.C., 1884), pp. 209–212. The best primary account of the movement toward international standard time can be found in *Proceedings of the American Metrological Society*, 1–5, 1873–88 (New York, 1889). A good summary appears in Ian R. Bartky, "The Adoption of Standard Time," in *Technology and Culture*, 30 (January 1989), pp. 25–56.

21. Howse, *Greenwich Time*, pp. 116–152. Quote from David S. Landes, *Revolution in Time: Clocks and the Making of the Modern World* (Cambridge, MA, 1983), p. 286. On time and World War I, see Kern, *The Culture of Time and Space*, pp. 259–289.

22. See Chapter Two and *Twenty-Ninth Annual Report of the Railroad Commissioners of the State of Connecticut* (Hartford, 1882).

23. Allen, *History of the Adoption of Standard Time*, pp. 12–13; *Traveler's Official Guide*, April 1882. Allen summarized the various plans briefly.

24. Allen, *History of the Adoption of Standard Time*, p. 13.

25. For an example, see Allen to Charles Pugh, Mar. 31, 1883 (AP, v. 1, p. 68).

26. Allen, *History of the Adoption of Standard Time*, pp. 16–21. The first map, badly faded, is now on file in the Division of Engineering and Industry, National Museum of American History, Washington, D.C. Numerous copies of the second remain in the Allen Papers. Allen paid special attention to asking the least possible exertion from the railroads. Most of the lines in the Northeast used New York, Boston, or Philadelphia time, and the differences between the three were relatively slight. The real problem came in east-west movement. By using existing breaks, Allen avoided most of these. Charles Dowd had come up with a similar plan in 1877, but by then he had lost the railroads' attention. See Charles F. Dowd, *The Railway Superintendent's Time Guide* (Saratoga Springs, NY, 1877), and *The Traveler's Railway Time Adjuster* (Saratoga Springs, NY, 1878).

27. *Proceedings of the American Railway Association*, I (1893), pp. 690–691.

28. See replies of M. E. Mathews (n.d.), J. J. Phano, Oct. 1, 1883, and L. F. Dutton, Sept. 12, 1883 (AP, Bx 1, fl. 3); George Crocker to Allen, Oct. 8, 1883 (AP, v. 2, p. 97); E. M. Reed to Allen, Sept. 3, 1883 (AP, v. 1, p. 127). Allen's reply to Reed, Sept. 4, 1883 (AP, v. 1, p. 153), and reply of J. D. Layng, Sept. 7, 1883 (original missing from AP—photograph of circular on file in the Division of Engineering and Industry, National Museum of American History, Washington, D.C.). Allen addressed most objections either in personal meetings or in correspondence which he did not preserve. A sample circular appears in Allen, *History of the Adoption of Standard Time*, pp. 19–21.

29. Allen to H. B. Ledyard, Oct. 4, 1883 (AP, v. I, pp. 292–293).

30. Allen to Connecticut Railroad Commissioners, April 18, 1883 (AP, v. I, p. 85). The legislature, however, had resolved not to take on any new business that session and would not hear the measure. See George P. Utley to Allen, May 2, 1883 (AP, v. I, p. 93). The question was decided automatically when New York City adopted the seventy-fifth meridian on Nov. 18.

31. W. H. Barnes to Allen, Sept. 8, 1883 (AP, v. I, p. 137); John Adams to Allen, Oct. 2, 1883 (AP, v. II, p. 68). Allen assured them that the observatories would follow the railroads, while privately taking care to see to it that they would. Allen to Barnes, Sept. 10, 1883 (AP, v. I, p. 170); Allen to Adams, Oct. 4, 1883 (AP, v. I, p. 299).

32. Abbe to Allen, Oct. 1, 1883 (AP, v. I, p. 272). Pickering favored standard time but believed that the demand for it should come from the public. See *Annual Reports of the Director of Harvard College Observatory, 1882–1883* (Boston, 1884), pp. 12–13.

33. Allen, *History of the Adoption of Standard Time*, pp. 23–24, 47–49. Edmands to Allen, Oct. 8, 1883 (AP, v. II, p. 29). Edmands to Allen, Nov. 6, 1883 (AP, v. IV, p. 32).

34. *Boston Evening Transcript*, Oct. 15, 1883; Oct. 29, 1883.

35. *Boston Globe*, Oct. 14, 1883; Nov. 17, 1883; *Daily Advertiser*, Oct. 25, 1883; City of Boston, *Report of the Committee on Standard Time*, Document 161 (Boston, 1883), pp. 2–7.

36. *Boston Daily Advertiser*, Nov. 3, 1883, p. 4.

37. Unidentified newspaper clipping, with letter dated Dec. 19, 1883, in AP, v. V, p. 64. See also *Boston Herald*, Jan. 9, 1884, p. 2.

38. Boston, *Report of the Committee on Standard Time*, pp. 5–7; *Boston Daily Advertiser*, Nov. 3, 1883, p. 4; Nov. 15, 1883.

39. *Proceedings of the ARA*, pp. 698–699, and Allen, *History of the Adoption of Standard Time*, pp. 58–59. Allen to J. Rowan, Oct. 6, 1883 (AP, v. I, p. 304).

40. Allen, *History of the Adoption of Standard Time*, pp. 24–25. The Convention met October 11, 1883. Nowhere does Allen or anyone else give a reason for choosing Nov. 18, except that Sundays usually saw less traffic, thus minimizing the potential for accidents.

41. Allen to J. M. Toucey, Oct. 4, 1883 (AP, v. I, p. 288).

42. Allen to Eckert, Sept. 14, 1883 (AP, v. I, p. 195).

43. Allen to Pugh, Sept. 28, 1883 (AP, v. I, p. 249); Pugh to Allen, Oct. 1, 1883 (AP, v. II, p. 64); Bates to Allen, Nov. 17, 1883 (AP, v. IV, p. 120).

44. See Allen, *History of the Adoption of Standard Time*, pp. 26–27, especially reprints of Mayor Edson's letter and the board of aldermen's

report on pp. 44–47; and the numerous articles in the New York newspapers the week before and after the change.

45. Baltimore and Ohio Railroad Co., *Circular No. 200*, Nov. 14, 1883. Photograph of circular on file in the Division of Engineering and Industry, National Museum of American History, Washington, D. C.

46. Quote from *St. Louis Globe-Democrat*, Nov. 18, 1883, p. 1; see also *New York Herald*, Nov. 19, 1883, p. 1; *New York Times*, Nov. 18, 1883, p. 3; *New York World*, Nov. 19, 1883, p. 1; *Chicago Tribune*, Nov. 19, 1883, p. 1; Pittsburgh *Dispatch*, Nov. 19, 1883, p. 2; and *Washington Post*, Nov. 18, 19, 1883, p. 1.

47. *St. Louis Globe-Democrat*, Nov. 19, 1883, p. 2; see also *New York Herald*, Nov. 19, 1883, p. 6; *New York Times*, Nov. 19, 1883, p. 5.

48. *New York Times*, Nov. 19, 1883, p. 5; *Chicago Tribune*, Nov. 14, 1883, p. 6. Western cities generally paid little or no attention to the change. San Francisco noted it with approval.

49. *Chicago Tribune*, Nov. 19, 1883, p. 1; *St. Louis Globe-Democrat*, Nov. 19, 1883, p. 2.

50. *Washington Post*, Nov. 18, 1883, p. 1.

51. *Indianapolis Sentinel*, Nov. 19, 1883, p. 8; *New York Times*, Nov. 19, 1883, p. 5.

52. *Chicago Tribune*, Nov. 14, 1883, p. 6, offers a detailed look at how cities obtained time before 1883. See also Nov. 19, p. 1; Nov. 20, p. 6.

53. *Bangor Daily Whig and Statesman*, Nov. 20, 1883, p. 1, Benedict Bros. to Allen, Nov. 19, 1883 (AP, v. IV, p. 133); *Atlanta Constitution*, Nov. 17, 1883; *St. Louis Post-Dispatch*, Nov. 19, 1883. See also *St. Louis Globe-Democrat*, Nov. 18, 1883.

54. *Atlanta Constitution*, Nov. 18, 1883.

55. Ibid., Nov. 20, 1883.

56. *Omaha Daily Republican*, Nov. 21, 1883, p. 4.

57. *New York Times*, Nov. 18, 1883, p. 3; *Pittsburgh Dispatch*, Nov. 19, 1883, p. 2; *St. Louis Globe-Democrat*, Nov. 18, 1883, p. 1; *Boston Daily Advertiser*, Nov. 14, 1883, p. 1.

58. Anyone looking up accounts of the invention of standard time would be surprised to find Dowd getting most of the credit. Thanks largely

to his own efforts, Dowd was often cited as the inspiration for the plan, while other accounts wrongly named F. A. P. Barnard. Annoyed, Allen threatened to resign from the American Metrological Society, and so Allen and Dowd were each invited to present that body with their own versions. They compromised, dividing the credit — Dowd for originating the idea of zones, Allen for working out the details and implementing the plan. Privately, Allen resented Dowd's intrusions. "The old crank," he wrote later, "would be amusing if he did not make himself so annoying" (AP, v. I, p. 654). Both men preserved the relevant papers to protect their claims. Ironically, Dowd died in 1904, struck by a train while crossing the railroad tracks near his home. The train was late and making up lost time. On the priority of invention, see Ian R. Bartky, "The Invention of Railroad Time," in *Railroad History*, 148 (Spring 1983), p. 22, notes 16 and 17.

59. *Harper's Weekly*, Dec. 29, 1883, p. 843.

60. *Proceedings of the American Railway Association*, 1 (1893), p. 703. On the International Meridian Conference and its effects, see Eviatar Zerubavel, "The Standardization of Time: A Sociohistorical Perspective," in *American Journal of Sociology*, 88 (July 1982), pp. 12–17; and Stephen Kern, *The Culture of Time and Space*, (New York, 1983), passim. The conference had no binding authority, and World Standard Time was only gradually adopted. Congressional attempts to formalize it by law failed until 1918.

61. William Graham Sumner, *What Social Classes Owe to Each Other* (1883), quoted in Richard Hofstadter, *Social Darwinism in American Thought* (1944; reprint ed. New York, 1955), p. 8.

62. *The Old Farmer's Almanac*, 1885 (Boston, 1884), p. 1; *Cincinnati Commercial Gazette*, Nov. 22, 1883, p. 4.

63. Though the case was occasionally cited as precedent over the next decades, a diligent search of both state and regional law reports and Holmes's private papers turned up no transcripts of his decision. The record of the case (*Clapp* vs. *Jenkins*, no. 1171) is now in the State Archives Building, University of Massachusetts, Boston. The case was a single Justice sitting of the Supreme Court and in such cases records were often not kept, according to the Clerk of the Suffolk County Courthouse. For reports of the case, see *Boston Daily*

Advertiser, Dec. 5, 1883, p. 2; *Boston Globe,* Dec. 5, 1883; *New York Times,* Dec. 5, 1883, p. 4.

64. *Bangor Daily Whig and Statesman,* Nov. 13, 1883, p. 3.

65. *Railway Age,* Nov. 29, 1883, p. 753; *Boston Journal* quoted in *Bangor Daily Whig and Statesman,* Nov. 15, 1883, p. 2.

66. *Railway Age,* Dec. 13, 1883, p. 787. *Railway Age* first called Cummings "A. Dogberry." Some of the few subsequent historical accounts of standard time missed the joke.

67. Arnold A. Lasker, "Standard Time Had a Rough Time in Bangor," in *Maine Historical Society Quarterly,* 24(2) (Fall 1984), pp. 276–277. Augusta *Journal* quoted in *New York Times,* Mar. 30, 1884, p. 14. See Allen's denial in the *Times,* April 1, 1884, p. 2, and a rebuttal from Maine, also in the *Times,* April 13, 1884, p. 2. See also Payson Tucker to Allen, Oct. 6, 1886 (AP, v. VI, no. 2A); Tucker to J. H. Manley, Oct. 4, 1886 (AP, v. VI, p. 1). Allen described his wooing of the Maine legislature in the *New York Tribune,* Jan. 1, 1887, p. 5. He referred to an "extensive correspondence" on the subject but did not preserve these letters.

68. See, for example, *Chicago Tribune,* Nov. 20, 1883, p. 6. Many other papers reprinted the story. On standard time in the District of Columbia, see *CR,* Mar. 13, 1884.

69. *New York Times,* Nov. 13, 1883, p. 3; *CR,* Mar. 10, 1884, pp. 1760–1763. A bill to fix a national time standard rose and fell very briefly in 1891.

70. Letter, response, and editorial in *Louisville Courier-Journal,* Nov. 22, 1883, p. 4. *New York Times,* June 28, 1884, p. 3; *Rochester German Ins. Co.* vs. *Peaslee,* June 13, 1905 (87 SW 1115). The case involved the time of expiration of an insurance contract.

71. *Wheeling Daily Register,* April 1, 1887, p. 4; Letter to Allen, Oct. 4, 1886 (AP, v. V, p. 161); *Pittsburgh Dispatch,* Dec. 28, 1886, p. 4; Jan. 1, 1887, p. 1. Thomas Schlereth tells me that as late as the 1950s, before the Vatican II reforms, Pittsburgh Catholics joked that the period of fast before taking Communion actually began at twelve-twenty by the clock, not midnight, since eastern time was twenty minutes ahead of the Pittsburgh meridian.

72. See the *Savannah Morning News,* Nov. 16, 1883, p. 2; Nov. 23, 1883, p. 4; Nov. 26, 1883, p. 3.

73. Ibid., Mar. 22, 1888, p. 8; *Augusta Chronicle,* April 30, 1888.

74. *Henderson* vs. *Reynolds,* Dec. 16, 1889 (10 SE 734).

75. *Indianapolis Daily Sentinel,* Nov. 21, 1883, p. 4; *Indianapolis Journal,* Nov. 21, 1883, p. 3.

76. *Charleston News and Courier,* Nov. 26, 1883, p. 1; *Natchez Daily Democrat,* Nov. 27, 1883, p. 1.

77. Quotes from *Cincinnati Commercial Gazette,* Nov. 24, 1883, p. 4; Nov. 22, 1883, p. 4; Nov. 18, 1883, p. 4; Feb. 20, 1890, p. 3.

78. Reprinted in the *Columbus Dispatch,* Nov. 27, 1883.

79. *Bellmont Chronicle,* Feb. 7, 1889, p. 1; *Wheeling Daily Register,* Feb. 22, 1889, p. 4; *New York Times,* Feb. 23, 1889, p. 1.

80. Letter to Allen from the Cleveland, Columbus, Cincinnati and Indianapolis Railway, Feb. 21, 1884 (AP, v. V, p. 106).

81. *Cincinnati Commercial Gazette,* Nov. 17, 1883, p. 7; Doris Chase Doane, *Time Changes in the U.S.A.* (San Francisco, 1973), p. 83; Henry Ford, *My Life and Work* (Garden City, NY, 1922), p. 24; Charles F. Dowd, "The Twenty-Four Hour Notation Leading to a Single Standard of Time and a Universal American Day" (2nd ed., n.d.) in the Dowd Papers, Manuscript Division, Boston Public Library, 19.9, no. 20. Dowd tried presenting this plan to the General Time Convention in 1884, but was denied a hearing.

82. Letter to Allen from the Cleveland, Columbus, Cincinnati and Indianapolis Railway, Feb. 21, 1884 (AP, v. V, p. 106); Mendenhall to Allen, Jan. 7, 1890 (AP, v. VI, p. 46); *Eleventh Annual Report of the Ohio Society of Surveyors and Civil Engineers* (Columbus, 1890), pp. 66–67; Ohio, *Journal of the House of Representatives,* Feb. 7, 1893, p. 222.

83. *Jones* vs. *German Ins. Co.,* Dec. 15, 1899 (81 NW 188).

84. See *Searles* vs. *Averhoff* (44 NW 873); *Rochester German Ins. Co.* vs. *Peaslee* (87 SW 1115); *Globe and Rutgers Fire Co.* vs. *Moffat,* NY, 1917 (154 FED 13). Additional examples include *Ex Parte Parker,* TX, 1895 (29 SW 480); *Texas T. & L. Co.* vs. *Hightower,* TX, 1906 (96 SW 1071); and *Walker* vs. *Terrel,* TX, 1916 (189 SW 75).

85. *State* vs. *Johnson*, MN, 1898 (77 NW 294); *Orvick* vs. *Cassleman*, ND, 1905 (105 NW 1105); *Salt Lake City* vs. *Robinson*, Utah, 1911 (116 PAC 442); *Goodman* vs. *Caledonian*, NY, 1917 (118 NE 524); *Bank of Fruitvale* vs. *Fidelity, etc., Co.*, CA, 1917 (35 CAL APP 666).

86. *Globe and Rutgers Fire Co.* vs. *Moffat*, NY, 1917 (154 FED 13); *The National Economist*, April 19, 1890, and W. Scott Morgan, *History of the Wheel and the Alliance and the Impending Revolution* (1891), p. 88.

87. *National Economist*, Mar. 14, 1889, p. 9. The article quoted ran as part of a weekly series under the theme "history repeats itself."

88. Nelson A. Dunning, *The Farmer's Alliance History and Agricultural Digest* (Washington, D.C., 1891), pp. 462 and 140, respectively.

89. Morgan, *History of the Wheel and the Alliance*, pp. 677 and 209, respectively; explanation of "the wheel," p. 62.

90. *Proceedings of the American Railway Association*, 1 (1893), pp. 732–733. Introduced in 1884, the regulations were offered as a model in the wake of the new time. They were not formally adopted by the Convention and may never have been applied on any railroad in precisely this form.

91. *Indianapolis Journal*, Nov. 16, 1883, p. 5.

92. *Cincinnati Commercial Gazette*, Nov. 18, 1883, p. 1.

93. Unidentified newspaper clipping dated Nov. 25, 1883, probably from the *New York* or *Boston Herald*, in AP, v. V, p. 12.

IV

Keep a Watch on Everybody

1. *New York Herald*, Nov. 19, 1883.

2. *New York Daily Tribune,* Dec. 10, 1895, p. 2. Thanks to Grace Palladino for this reference.

3. Lawrence W. Levine, *Highbrow/Lowbrow. The Emergence of Cultural Hierarchy in America* (Cambridge, MA, 1988), pp. 171–200; *Boston Daily Advertiser,* Jan. 1, 1884, p. 4.

4. *Music*, 2 (July 1892), pp. 253–254. Thanks to Lawrence W. Levine for this reference.

5. *McGuffey's Fifth Eclectic Reader* (1881; reprint ed. New York, 1920), pp. 161–163.

6. *McGuffey's Third Eclectic Reader* (1881; reprint ed. New York, 1920), pp. 72–74.

7. *Harper's New Monthly Magazine*, 68 (January 1884), p. 327. Warner never mentioned standard time or even clocks—he focused his complaints on the larger divisions of months, days, and years. But its appearance less than two months after the invention of standard time suggests a connection.

8. George M. Beard, *American Nervousness: Its Causes and Consequences* (1881; reprint ed. (New York, 1972), pp. 104–105. As an alternative to the public world of time pressure, Beard favored "rest cures" that isolated the patient from time constraints.

9. Late nineteenth- and very early twentieth-century advertising is notoriously hard to find, since ads were often cut out of magazines before binding for library storage. Regrettably little survives today. I have relied heavily on a few periodicals and on the outstanding collections of advertisements and trade catalogues in the Business Americana Collection, Archives Center, National Museum of American History (BAC); and the files of the Section of Mechanisms, Division of Engineering and Industry, also at the Museum of American History (DEI).

10. See *Main Features of P. H. Dudley's Electrically Controlled System of Railroad and City Time Service* (New York, 1882), and another example, *The Electric Signal and Time Register Co.* (New York, 1882), pamphlets on file in BAC, "Clocks and Watches," Bx 1. Dudley drew his signals from the Washburn Observatory in Wisconsin. See also *New York Tribune*, Feb. 7, 1885, p. 8, "Time Furnished Like Gas or Water," for another example. As Chapter Two related, the Naval Observatory took pains to get the Signal Bureau's competing time service suspended by the Secretary of the Navy. The Secretary's order made the Naval Observatory the only federally sanctioned source of time. Again, standard time, and the railroad meridians, never received any official government endorsement. Theoretically,

"Naval Observatory time" meant time by the meridian of Washington, D.C., not the seventy-fifth meridian, the eastern zone standard.

11. Application for Time Service by Standard Yarn Co. of Oswego, New York, April 10, 1889, in the Western Union Archives, Archives Center, National Museum of American History (BAC). The Self-Winding Clock Company seems to have held the position of exclusive hardware supplier for the Western Union time service. Applications for the service admitted that other clocks, properly prepared to receive the time signal, might be used. But they discouraged substitutions.

12. *Jeweler's Circular and Horological Review,* 21 (February 1890), p. 91. Hiero Taylor, "The United States Government System of Observatory Time," in *What Is Standard Time?* promotional pamphlet of the Self-Winding Clock Company, in DEI.

13. *Scientific American,* 69 (July 29, 1893), p. 69. The service was available for home use but cost too much for most homes. The time signals were relatively inexpensive, but the clocks themselves were not.

14. Catalogue of the Electric Signal Clock Company, 1891, p. 5. In BAC, "Clocks and Watches," Bx 3. The Autocrat was a master clock that could be made to ring bells automatically.

15. Catalogue of the Blodgett Signal Clock Company, 1896, p. 2. Photocopy in National Association of Watch and Clock Collectors (NAWCC) Library, Columbia, PA.

16. Catalogue of the Electric Signal Clock Company, p. 12.

17. Catalogue of the Blodgett Signal Clock Company, p. 2.

18. Ibid., p. 1.

19. For an intriguing and important discussion of the relationship between machines, efficiency, and individuality in this period, see Walter Benn Michaels, "An American Tragedy, or the Promise of American Life," in *Representations,* 25 (Winter 1989), pp. 71–98. Michaels argues that only standardization and machine models made it possible for individuals to find a niche, to free themselves from the organic ties of sentiment and tradition and take their place in world where their individual attributes would be studied and utilized

to the full. Standardization, in other words, harmonizes the notion of the individual with the realities of machine culture.

20. Neil McKendrick, "Josiah Wedgwood and Factory Discipline," in *The Historical Journal*, IV (1961), pp. 30–55; A. G. Bromley, "Charles Babbage and the Invention of Workmen's Time Recorders," in *Antiquarian Horology*, 13 (September 1982), pp. 442–447. The base of the lamp reads "Pat. Applied For," but there is no patent number or record of the inventor's name. Consequently, it is difficult to tell how widely the device was used.

21. Alan Sayles, "A Short History of Dial Recorders," in *Bulletin of the National Association of Clock and Watch Collectors*, 9 (1961), pp. 941–942. In Bundy's system the keys were kept on a board or in a cabinet next to the clock.

22. *The Bundy Time Recorder*, n.d. (ca. 1893), pamphlet on file in BAC, "Watches and Clocks," Bx 1. Later models usually included windows, to allow the workers to see if their time had been recorded properly.

23. 1914 Catalogue of the International Time Recording Company (ITRC), p. 7; letter from S. H. Chamberlain to the Peace Dale Manufacturing Company, Feb. 1, 1913; International Time Recording Company, "Time: The Gateway to a Better Business," 1916, p. 4. All in DEI.

24. 1914 Catalogue of the ITRC, p. 9. In DEI.

25. *Factory: the Magazine of Management*, 6 (January 1911), p. 6. See also Adrian Forty, *Objects of Desire: Design and Society from Wedgwood to IBM* (New York, 1986), p. 122.

26. 1918 catalogue of the ITRC, p. 1. In DEI. Recall the description of the Lowell Mills that ended Chapter Two. Here, ending the process begun at Lowell, the mechanical clock has completely replaced paternal authority, and rebuilt the bonds between employer and employee.

27. 1914 catalogue of the ITRC, p. 9. In DEI. Samuel Haber, *Efficiency and Uplift: Scientific Management in the Progressive Era* (reprint ed. Chicago, 1973), p. ix.

28. See Haber, *Efficiency and Uplift*, passim.

29. On the tension between individual and system in scientific management and the Progressive era, see Michaels, "An American Tragedy," pp. 71–98. On Taylor as a system builder, see Thomas P. Hughes, *American Genesis: A Century of Invention and Technological Enthusiasm* (New York, 1989), pp. 184–248.

30. Frank B. Copley, *Frederick W. Taylor, Father of Scientific Management* (New York, 1923), p. 52.

31. On Taylor and his disciples, see Copley's largely uncritical *Frederick W. Taylor*; Haber, *Efficiency and Uplift*; Daniel Nelson, *Frederick W. Taylor and the Rise of Scientific Management* (Madison, WI, 1980); Hugh Aitken, *Taylorism at the Watertown Arsenal: Scientific Management in Action, 1908–1915* (Cambridge, MA, 1960); Milton Nadworny, *Scientific Management and the Unions* (Cambridge, MA, 1955); and the discussions of Taylor in Daniel T. Rodgers, *The Work Ethic in Industrial America, 1850–1920* (Chicago, 1974). On rationalization in the home, see Catherine Beecher, *A Treatise on Domestic Economy* (1841; reprint ed. New York, 1970), p. 42 and the discussion of Beecher in Chapter One. Taylor's obsession with work came from his mother, not his father, who was said to have been "born retired." Women seem to have borne most of the burden of socializing their children to hard and efficient work. Dolores Hayden's *The Grand Domestic Revolution: A History of Feminist Designs for American Homes, Neighborhoods, and Cities* (Cambridge, MA, 1982) explores the long-standing emphasis, among American feminist reformers, on conservation of time and motion. See discussion of time in the home below.

32. Frederick W. Taylor, "Shop Management," in *Transactions of the American Society of Mechanical Engineers*, 24 (1902–03), pp. 1391–1406 (reprinted, slightly revised, as *Shop Management* [New York, 1911], pp. 100–122). See also Aitken, *Taylorism at the Watertown Arsenal*, pp. 13–34.

33. Hughes, *American Genesis*, pp. 184–248; Aitken, *Taylorism at the Watertown Arsenal*, pp. 28–29, 97; Haber, *Efficiency and Uplift*, p. 24; Taylor, *Shop Management*, pp. 102–103; *Principles of Scientific Management* (New York, 1911), pp. 124–125; Copley, *Frederick W. Taylor*, vol. 1, pp. 358–370.

34. Copley, following one of Taylor's several versions, put the date as 1883, while Nelson's more careful review disclosed several alternative dates. See Nelson, *Frederick W. Taylor*, p. 41. As Nelson and others have pointed out, Taylor's memory was often conveniently faulty.

35. U.S. Congress, *Special Committee to Investigate the Taylor and Other Systems of Shop Management*, 3 vols. (Washington D.C., 1912). Testimonies of Orrin Cheney, vol. 1, p. 31; Richard Stackhouse, vol. 1, p. 295, 308–309; W. T. Probert, vol. 1, p. 757; and Olaf Nelson, vol. 1, p. 507. See also Rodgers, *The Work Ethic in Industrial America*, pp. 166–168.

36. Aitken, *Taylorism at the Watertown Arsenal*, p. 72. *Special Committee to Investigate the Taylor and Other Systems of Shop Management*, testimony of Joseph R. Cooney, vol. 1, 255.

37. Taylor, *Shop Management*, pp. 153–154.

38. A concise account is given in Aitken, *Taylorism at the Watertown Arsenal*, pp. 144–150. See also *Special Committee to Investigate the Taylor and Other Systems of Shop Management*, vol. 3, pp. 1888–1891. The molders misunderstood how Taylor's time was determined, and assumed that Taylor's man had simply lied. Twenty-four minutes represented the ideal time—under Taylor's standard formulas, premium pay rates would begin at any time under forty minutes.

39. *Special Committee to Investigate the Taylor and Other Systems of Shop Management*, testimony of Olaf Nelson, vol. 1, p. 508; testimony of John R. O'Leary, vol. 1, p. 18.

40. E. P. Thompson, "Time, Work-Discipline, and Industrial Capitalism," in *Past and Present*, 38 (December 1967), p. 86.

41. *CR*, Feb. 23, 1915, p. 4352.

42. It is sometimes claimed that officers were issued watches for the first time in the Civil War, but I have found no evidence to support this. It is true, however, that soldiers attached great prestige to owning a watch and that watch trading seemed to have been a popular pastime. On Waltham, see Charles W. Moore, *Timing a Century: The History of the Waltham Watch Company* (Cambridge, MA, 1945), passim. On the American watch industry, see Michael C. Harrold, *American Watchmaking: A Technical History of the American Watch Industry, 1850–1930* (1981), a supplement to the *Bulletin of the National Association of Watch and Clock Collectors*, 14 (Spring 1984), and the

concise account of David S. Landes, *Revolution in Time: Clocks and the Making of the Modern World* (Cambridge, MA, 1983), pp. 290–320.

43. Quoted in Landes, *Revolution in Time*, p. 307.

44. Landes, *Revolution in Time*, p. 318; Moore, *Timing a Century*, p. 53; National (Elgin) Watch Company, *Illustrated Almanac, 1871* (New York, 1870).

45. John G. Cawelti, *Apostles of the Self-Made Man: Changing Concepts of Success in America* (New York, 1965), p. 118.

46. Henry Ford, *My Life and Work* (Garden City, NY, 1925), p. 24. Watch clubs described in Boris Emmet and John E. Jeuck, *Catalogues and Counters: A History of Sears, Roebuck and Company* (New York, 1950), p. 28. For a rural example, see ad of J. C. Saunders and Co. in the *Bonham* (Texas) *News*, July 12, 1889–"A man working for a salary has no excuse for not owning a watch," the ad claimed.

47. Waterbury Watch Company, *Tick Tock Tick and Other Rhymes*, dated "ca. 188?" in Library of Congress.

48. Waterbury Watch Company, *Round and Round the Wheels Go Round* (New York, 1887), pamphlet in BAC, Bx 6, "Waterbury Watch Co." folder.

49. *New Haven Evening Register*, Mar. 1, 1887, p. 4.

50. *Round and Round the Wheels Go Round* (BAC, Bx 6, "Waterbury Watch Co." folder).

51. Harold Wentworth and Stuart Berg Flexner, eds., *Dictionary of American Slang* (New York, 1960), p. 518; and Flexner, *I Hear America Talking: An Illustrated History of American Words and Phrases* (New York, 1976). Thanks to Tom Schlereth for the information about Phi Beta Kappa, and to Mark Meigs for help with these ideas.

52. Reverend Alfred Taylor, *The Watch and the Clock, Number Thirty-Seven of the Home College Series* (New York, 1883), p. 13; U.S. Bureau of Standards, *Circular No. 51: Measurement of Time and Test of Timepieces* (Washington, D.C., 1914), pp. 18–19.

53. *The Waterbury*, 1 (October 1887), p. 62; *Harper's New Monthly Magazine*, 39 (July 1869), pp. 169–182.

54. *The Waterbury*, 2 (August 1888), p. 28.

55. Waterbury Watch Company, *Keep a Watch on Everybody* (New York, 1887) (BAC, Bx 6, "Waterbury Watch Co." folder).

56. National (Elgin) Watch Company, *Illustrated Almanac, 1875* (New York, 1874).

57. Lewis Mumford, *Technics and Civilization* (1934; reprint ed. New York, 1963), p. 14.

58. *The Waterbury*, 2 (December 1888), p. 91; 2 (January 1889).

59. *The Waterbury*, (February 1888), p. 121; *Collier's*, 44 (November 6, 1909), p. 5.

60. *Cleveland Gazette*, Nov. 3, 1883. The watch is billed ahead of the fifty dollars, though probably worth at the very most about half that. The ad reinforces the value of the watch as a universal status object, and the reference to "stem-winding" associates the watch with the aggressive man on the rise.

61. Boris Emmet and John E. Jeuck, *Catalogues and Counters: A History of Sears, Roebuck and Company* (New York, 1950), pp. 25–35.

62. Sears, Roebuck and Co., *1897 Catalogue* (reprint ed. New York, 1968), p. 369; *National Economist*, Oct. 18, 1890, p. 83. The ad is for Sears as "The Warren [Sears's middle name] Company."

63. Sears, Roebuck and Co., *1897 Catalogue*, p. 369.

64. Emmet and Jeuck, *Catalogues and Counters*, p. 33. Montgomery Ward and Co., *1895 Catalogue and Buyers Guide* (reprint ed. New York, 1969); Sears, Roebuck and Co., *1897 Catalogue*. Though the mail-order catalogues emphasized pastoral scenes more than more urban icons, it would be wrong to ignore the general popularity of rural scenes in other decorative arts of the period—in jewelry, for example. It might easily be argued that the general popularity of rural scenes as a decorative motif indicates a widespread ambivalence about machine production and industrialization. I only hope to call attention to the particularly striking juxtaposition of pastoralism with the pocketwatch, an unmistakably industrial machine, and the ambivalence about time and technology it suggests.

65. Ward and Co., *1895 Catalogue and Buyers Guide*, p. 144.

66. Dolores Hayden, *The Grand Domestic Revolution: A History of Feminist Designs for American Homes, Neighborhoods, and Cities* (Cambridge, MA,

1982); Beecher, *A Treatise on Domestic Economy* (1841; reprint ed. New York, 1970), p. 171.

67. See Susan Strasser, *Never Done: A History of American Housework* (New York, 1982); and Ruth Schwartz Cowan, *More Work for Mother: The Ironies of Household Technology from the Open Hearth to the Microwave* (New York, 1983); Daniel Rodgers, *The Work Ethic in Industrial America, 1850–1920* (Chicago, 1978), pp. 182–209.

68. Cowan, *More Work for Mother*, pp. 12–13.

69. On the ideology of separate spheres, see Katherine Kish Sklar, *Catherine Beecher: A Study in American Domesticity* (New Haven, 1973); Nancy Cott, *The Bonds of Womanhood* (New York, 1977); and Strasser, *Never Done*, pp. 181–201.

70. Catherine Beecher and Harriet Beecher Stowe, *The American Woman's Home* (New York, 1869), p. 372; Todd S. Goodholme, *Goodholme's Domestic Encyclopedia* (New York, 1878); Frank and Marion Stockton, *The Home: Where it Should Be and What to Put In It* (New York, 1873), pp. 76–77.

71. Hayden, *The Grand Domestic Revolution*, p. 128.

72. Quote from Maria Parloa, *Home Economics* (New York, 1906), p. 48. On cooking and recipes, see Laura Shapiro, *Perfection Salad: Women and Cooking at the Turn of the Century* (New York, 1986); Sidney Morse, *Household Discoveries* (New York, 1908), pp. 767–770; Ruth A. Beezly and Annie Gregory, *The National Course in Home Economics* (New York, 1917), frontispiece.

73. Quote from Thetta Quay Franks, *The Margin of Happiness* (New York, 1917), pp. 79–80. On scientific management in the home, see Cowan, *More Work for Mother*, ch. 6; Strasser, *Never Done*, pp. 213–219; Forty, *Objects of Desire*, pp. 114–120. For examples, see Martha Bensley Bruere and Robert W. Breure, *Increasing Home Efficiency* (New York, 1912), and Christine Frederick, *Household Engineering: Scientific Management in the Home* (Chicago, 1920).

74. Rodgers, *The Work Ethic in Industrial America, 1850–1920*, pp. 94–99.

75. Mrs. Henry Ward Beecher (Eunice White Bullard), *The Home: How to Make It and Keep It* (Minneapolis, 1883).

76. Elsie DeWolf, *The House in Good Taste* (New York, 1913), p. 5. The extreme example of personality expressed through consumer goods

would of course be Dreiser's Sister Carrie, the woman who literally makes herself over in the consumer marketplace. For a discussion of personality in home furnishings, see Forty, *Objects of Desire*, pp. 104–115 esp.

77. See Warren Susman, " 'Personality' and the Making of Twentieth Century Culture," in *Culture as History: The Transformation of American Society in the Twentieth Century* (New York, 1984), pp. 271–285.

78. De Wolf, *The House in Good Taste*, pp. 3, 300–306.

79. Examples of clock design drawn from trade catalogues on file in DEI, including Waterbury Clock Company, *Catalogue Numbers 154* (1908–09) and *156* (1910–11); F. Kroeber Clock Company, *Illustrated Catalogue, 1888–1889* (reprint ed. Kansas City, 1983); Ansonia Clock Company, *Catalogue of the Ansonia Clock Company, 1886–1887* (reprint ed. Ironton, MO, 1978); *1904–1905* (New York, 1904), and *1914* (reprint ed. Benicia, CA, 1983); and the survey of American clock design in Anita Schorsch, *The Warner Collector's Guide to American Clocks* (New York, 1981).

80. De Wolf, *The House in Good Taste*, p. 3. On the splintering of private time and its connection to public standardization, see Stephen Kern, *The Culture of Time and Space* (Cambridge, MA, 1983), pp. 10–30.

81. *Saturday Evening Post,* Nov. 4, 1916, p. 40.

82. Ibid., April 15, 1916, p. 39.

83. Michaels, "An American Tragedy," in *Representations,* 25 (Winter 1989), pp. 71–98. Michaels argues that in the early twentieth century individuality could only exist within mechanical social systems—that becoming a cog in Taylor's machine, or wearing standardized clothing styles, granted the individual a "place," a degree of individuality and autonomy not found in "organic" society.

84. *Saturday Evening Post,* Oct. 7, 1916, p. 39.

85. Ibid., Oct. 14, 1916, p. 47.

86. Ibid., Jan. 27, 1912. Thanks to Charles McGovern for this reference.

87. *Ladies' Home Journal* (December 1915), p. 74. Thanks to Miriam Formanek Brunell for this reference.

V

Therbligs and Hieroglyphs

1. Robert Bartlett Haas, *Muybridge: Man in Motion* (Berkeley, CA, 1976); Kevin McDonnel, *Eadweard Muybridge: The Man Who Invented the Moving Picture* (Boston, 1972); Gordon Hendricks, *Eadweard Muybridge: The Father of the Moving Picture* (New York, 1975): Craig Zabel, "Capturing Time: Muybridge and the Nineteenth Century," in David Robertson, ed., *The Art and Science of Eadweard Muybridge* (Carlisle, PA, 1985); and Robert Sklar, *Movie-Made America: A Cultural History of American Movies* (New York, 1975), pp. 3–9.

2. On the relationship between American art and scientific management, see Cecelia Tichi, *Shifting Gears: Technology, Literature, Culture in Modernist America* (Chapel Hill, NC, 1987); Stephen Kern, *The Culture of Time and Space* (Cambridge, MA, 1983); and Donald M. Lowe, *History of Bourgeois Perception* (Chicago, 1982).

3. Lary May, *Screening Out the Past: The Birth of Mass Culture and the Motion Picture Industry* (Chicago, 1983), p. 25; David A. Cook, *A History of Narrative Film* (New York, 1981), pp. 61–62.

4. Cook, *History of Narrative Film*, pp. 7–8; Gordon Hendricks, "The Kinetoscope: Fall Motion Picture Production," in John L. Fell, ed., *Film Before Griffith* (Berkeley, CA, 1983), pp. 13–21.

5. May, *Screening Out the Past*, pp. 36–37. On early American film makers, see Charles John Musser, "Before the Nickelodeon: Edwin S. Porter and the Edison Manufacturing Company" (Ph.D. diss., NYU, 1986), and his forthcoming book; John L. Fell, *Film and the Narrative Tradition* (1974; reprint ed. Berkeley, CA, 1986); David Bordwell, Janet Staiger, and Kristin Thompson, *The Classical Hollywood Cinema: Film Style and Mode of Production to 1960* (New York, 1985); and Kemp R. Niver, *The First Twenty Years* (Los Angeles, 1968).

6. Sklar, *Movie-Made America*, p. 17. Movie exhibitors apparently felt free to alter the films. Musser, "Before the Nickelodeon," calls the exhibitors of the period "co-creators" of the audience's experience, and sees their construction of film programs as an antecedent of movie editing.

7. *The Nation*, Aug. 28, 1913, quoted in May, *Screening Out the Past*, pp. 35–36. See also Sklar, *Movie-Made America*, pp. 18–19.

8. *The Independent*, Feb. 6, 1908, pp. 307–308.

9. *Harper's Weekly*, 51 (Aug. 24, 1907), p. 573; *The Independent*, 69 (Sept. 29, 1910), p. 715. See also C. Francis Jenkins, *Animated Pictures* (Washington, D.C., 1898), pp. 102–106.

10. *Survey*, June 5, 1909, pp. 355–365; *Harper's Weekly*, 51 (Aug. 24, 1907), p. 573.

11. *The Dial*, 57 (Sept. 1, 1914), p. 127.

12. *Scientific American*, 100 (June 26, 1909), p. 476.

13. Georges Méliès, *The Clockmaker's Dream* (New York & Paris, 1904), Paperprint Collection, Motion Picture, Television and Sound Division, Library of Congress. The film had a very smooth continuity; it had a narrative flow, but no real story. Méliès's films were so often copied that in 1903 he finally opened a New York office to control distribution and deter felonious American rivals. On film and magic, see Eric Barnouw, *The Magician and the Cinema* (New York, 1981).

14. For a general description of film structure and content in this period, see John L. Fell, "Motive, Mischief, and Melodrama: The State of Film Narrative in 1907," in Fell, ed., *Film Before Griffith* (Berkeley, CA, 1983), pp. 272–285.

15. *Harper's Weekly*, 57 (Jan. 18, 1913), p. 20.

16. Cook, *History of Narrative Film*, pp. 32–33. For examples of recent scholarship, see Musser, "Before the Nickelodeon," and the various essays in Fell, ed., *Film Before Griffith*.

17. *Survey*, 22 (June 5, 1909), pp. 355–365. See also *Harper's Weekly*, 54 (July 30, 1910), pp. 12–13; and Roy Rosenzweig, *Eight Hours for What We Will: Workers and Leisure in an Industrial City, 1870–1920* (Cambridge, MA, 1985). Lawrence W. Levine, *Highbrow/Lowbrow. The Emergence of Cultural Hierarchy in America* (Cambridge, MA, 1988), points out that this kind of audience behavior was disappearing.

18. Rosenzweig, *Eight Hours for What We Will*, p. 201; *Harper's Weekly*, 61 (Dec. 11, 1915), p. 574.

19. *Scientific American*, 112 (May 15, 1915), pp. 454–455.

20. *American Magazine*, 76 (September 1913), p. 60; *Harper's Weekly*, 57 (January 18, 1913), p. 20.

21. *Survey*, 32 (June 27, 1914), p. 337. Detailed discussions of movie censorship can be found in May, *Screening Out the Past*, ch. 3; Sklar, *Movie-Made America*; Rosenzweig, *Eight Hours for What We Will*; and Kathy L. Peiss, *Cheap Amusements: Working Women and Leisure in Turn-of-the-Century New York* (Philadelphia, 1986).

22. Conflicting prints of the film survive. Though the version described is now generally regarded as the "original" as Porter released it, some believe the repeated action of the last scenes was often recut in a more modern form. See Musser, "Before the Nickelodeon," pp. 265–272; and Cook, *A History of Narrative Film*, pp. 18–20.

23. André Gaudreault, "Temporality and Narrativity in the Early Cinema," in Fell, ed., *Film Before Griffith*, pp. 316–319.

24. David Wark Griffith, "The Fatal Hour" (1908), Paperprint Collection, Motion Picture, Television and Sound Division, Library of Congress. Thanks to Charles Musser for bringing the film to my attention.

25. *The Independent*, 69 (Sept. 29, 1910), p. 714.

26. *Harper's Weekly*, 61 (Dec. 11, 1915), p. 573.

27. Between 1910 and 1919, New York newspapers give a mixed picture. Major features or special presentations of news started at precise times, but many theaters simply listed the names of new films, or major stars, with no show times. Obviously, the transition to feature films proceeded slowly; the one-reel format lasted into the twenties in some cases. On the other hand, many more films were produced and shown than advertised. Most theaters, counting on neighborhood patronage, may simply have posted starting times outside and assumed that customers would note the times on their way by.

28. *American City*, 7 (Sept. 12, 1912), pp. 234–38; *Literary Digest*, Aug. 5, 1916, p. 508.

29. *The Policy and Standards of the National Board of Censorship of Motion Pictures* (New York, 1914), pp. 10–11, passim. Thanks to Kathy Piess for loaning me a copy of this document.

30. *Current Opinion,* 63 (October 1917), p. 251. For examples of different ad formats, see *New York Tribune,* Nov. 13, 1916, p. 3; and Nov. 13, 1916, p. 7.

31. See Bordwell, Staiger, and Thompson, *The Classical Hollywood Cinema,* pt. 3.

32. *Survey,* 23 (Aug. 27, 1910), p. 740. See also Randolph Bourne's critique of a similar film in *New Republic,* 3 (July 3, 1915), p. 233.

33. *Good Housekeeping* (Aug. 10, 1910), pp. 184–185.

34. *Lippincott's,* 84 (October 1909), p. 455.

35. Seymour Chatman, "What Novels Can Do that Films Can't (and Vice Versa)," in W. J. T. Mitchell, ed., *On Narrative* (Chicago, 1981), p. 126 and passim, makes some related points.

36. Noël Burch, "Porter, or Ambivalence," in *Screen,* 19 (Winter 1978–79), p. 104.

37. Hugo Münsterberg, *The Photoplay: A Psycholocial Study,* 1916 (quoted in Dover reprint ed., retitled *The Film: A Psychological Study,* (New York, 1970), pp. 74, 95.

38. *Lippincott's,* 84 (October 1909), p. 455.

39. Münsterberg, *The Photoplay,* p. ix; *Psychology and Industrial Efficiency* (Boston, 1913), pp. 37–56; *American Problems: From the Point of View of a Psychologist* (1910; reprint ed. New York, 1969), pp. 25–43.

40. Münsterberg, *The Photoplay,* p. 94.

41. *Scientific American,* 56 (Jan. 13, 1917), p. 56.

42. *Harper's Weekly,* 58 (June 6, 1914), p. 21.

43. Leslie T. Peacocke, *Hints on Photoplay Writing* (Chicago, 1916), pp. 21, 28–29; *Colliers,* 52 (Nov. 1, 1913), pp. 24–26; Frederick A. Talbot, *Moving Pictures: How They Are Made and Worked* (Philadelphia, 1914), pp. 146–147; *Moving Picture World* quoted in Janet Staiger, "Dividing Labor for Production Control: Thomas Ince and the Rise of the Studio System," in *Cinema Journal,* vol. 18, no. 2 (Spring 1979), p. 19. See also *Harper's Weekly,* 56 (Sept. 7, 1912), p. 13.

44. *Saturday Evening Post,* 188 (May 13, 1916), p. 14. See also *New York Times,* Sept. 28, 1919, sec. IV, p. 5.

45. On Ince, see Janet Staiger, "Dividing Labor for Production Control: Thomas Ince and the Rise of the Studio System," in *Cinema Journal*, vol. 18, no. 2 (Spring 1979), pp. 16–25; and her chapters on "The Hollywood Mode of Production to 1930," in Bordwell, Staiger, and Thompson, *The Classical Hollywood Cinema*, pp. 85–142. See also the account of Ince's career in Lewis Jacobs, *The Rise of the American Film* (New York, 1939), pp. 204–206. Jacobs claims Ince stamped his scripts "produce this exactly as written!" in order, like Taylor, to disempower his employees and eliminate their input into the manufacturing process.

46. Bordwell, Staiger, and Thompson, *The Classical Hollywood Cinema*, p. 162.

47. *Survey*, Sept. 13, 1913, p. 724.

48. *Survey*, Sept. 6, 1913, pp. 684, 691.

49. Frank B. and L. M. Gilbreth, *Applied Motion Study* (1916; reprint ed. Easton, PA 1973), p. 203. The Gilbreths' "One Best Way" turned into a sort of Benthamite calculus of good. An innovation's worth, they insisted, depended on the number of "happiness minutes" it produced. On the lives of the Gilbreths, see Edna Yost, *Frank and Lillian Gilbreth: Partners for Life* (New Brunswick, NJ, 1949).

50. *Current Opinion*, 55 (September 1913), p. 206.

51. L. M. Gilbreth, *The Psychology of Management* (New York, 1914); Münsterberg, *American Problems*, pp. 25–33; and *Psychology and Industrial Efficiency*, p. 164. Münsterberg visited the Gilbreths' laboratory and cited their work in the latter.

52. On Gilbreth's relationship with Taylor and his disciples, see Daniel Nelson, *Frederick W. Taylor and the Rise of Scientific Management* (Madison, WI, 1980), pp. 131–136; Samuel Haber, *Efficiency and Uplift: Scientific Management in the Progressive Era, 1820–1920* (Chicago, 1973), pp. 37–42; and Milton Nadworney, "Frederick Taylor and Frank Gilbreth: Competition in Scientific Management," in *Business History Review*, XXXI (Spring 1957), pp. 23–34.

53. Gilbreths, *Applied Motion Study*, pp. 65–70.

54. Frank and Lillian Gilbreth, "Abstract of the Effect of Auto Micro-Motion Study on Educational and Industrial Methods" (Nov. 23, 1916), in a set of microfilm reels, "Frank B. Gilbreth, Selected

Papers, ca. 1910–1924," from the Gilbreth Lilbrary, Purdue University Libraries, Lafayette, Indiana (hereafter referred to as GP), Reel 1.

55. Gilbreths, *Applied Motion Study*, p. 119, See also Frank and Lillian Gilbreth, *Fatigue Study* (1916; reprint ed. Easton, PA, 1973), p. 126.

56. Adapted from chart in GP, Reel 1.

57. Vachel Lindsay, *The Art of the Moving Picture* (1915; reprint ed. New York, 1970), pp. 172–173. In his 1922 revision of the book, Lindsay commented on America's "hieroglyphic civilization." Advertisements, cartoons, photography, and the movies "make us into a hieroglyphic civilization far nearer to Egypt than England," pp. 21–22.

58. Quotes from Lindsay, *Art of the Moving Picture*, pp. 65–66; Gilbreths, "Abstract of the Effect of Auto Micro-Motion Study," p. 4 (GP, Reel 1). Thanks to Michael Rogin for help with these points.

59. *The Independent*, 69 (Sept. 29, 1910), p. 713.

60. *Scientific American Supplement*, 80 (May 22, 1915), p. 323; *The Bookman*, 47 (May 1918), p. 238.

61. Lindsay, *Art of the Moving Picture*, p. 54.

62. Recent scholarship on Griffith downplays both his originality and the idea that he "invented" the modern film vocabulary. For examples, see Michael P. Rogin, "The Sword Became a Flashing Vision: D. W. Griffith's *The Birth of A Nation*," in *Ronald Reagan, The Movie, and Other Episodes in Political Demonology* (Berkeley, CA, 1987); Bordwell, et al., *The Classical Hollywood Cinema*, part 3, passim; Cook, *A History of Narrative Film*, pp. 59–92; and Tom Gunning, "Weaving a Narrative: Style and Economic Background in Griffith's Biograph Films," in *Quarterly Review of Film Studies*, 6 (Winter 1981), pp. 12–25.

63. Rogin, *Ronald Reagan, The Movie*, pp. 199–200. I am greatly indebted to Michael Rogin for his many provocative discussions of Griffith and American film.

64. Lindsay, *Art of the Moving Picture*, p. 49.

65. *St. Louis Mirror* quoted in *Current Opinion*, 64 (May 1918), p. 332.

66. Wilson quoted in Rogin, *Ronald Reagan, The Movie*, p. 192.

67. Woodrow Wilson, "After-Dinner Remarks in New York to the Motion Picture Board of Trade," Jan. 27, 1916, reprinted in Arthur Link, ed., *The Papers of Woodrow Wilson* (Princeton, NJ, 1985), vol. 36, pp. 16–17.

68. *Current Opinion*, 63 (December 1917), pp. 346–347. Despite his optimistic words, Griffith's romantic vision of war collapsed after he visited the trenches. *Hearts of the World* made money, thanks to the era's superheated patriotism. But the film lacked the narrative continuity and simulacrum of history that distinguished *Birth*. Griffith was unable to connect the atrocities he had seen in Europe with any grand, romantic principles of crusade, and the film's narrative trajectory led toward disgust, cynicism, and disillusion. Wilson apparently found the film repellent, and after seeing it he severed his friendship with Griffith permanently. See Russell Merritt, "D. W. Griffith Directs the Great War: The Making of Hearts of the World," in *Quarterly Review of Film Studies* (Winter 1981), pp. 45–65.

69. Wilson quoted in Rogin, *Ronald Reagan, The Movie*, pp. 97, 94.

70. Larry Wayne Ward, *The Motion Picture Goes to War: The U.S. Government Film Effort During World War One* (Ann Arbor, MI, 1985), pp. 17–18; Terry Ramsaye, *A Million and One Nights: A History of the Motion Picture Through 1925* (New York, 1926), p. 728.

71. Münsterberg, *The Photoplay*, p. x.

72. Ward, *The Motion Picture Goes to War*, pp. 34–43. The films apparently do not survive.

73. On World War I motion pictures generally, see the detailed accounts in Ward, *The Motion Picture Goes to War*, pp. 10–19; David H. Mould, *American Newsfilm, 1914–1919: The Underexposed War* (New York, 1983); Michael Isenberg, *War on Film: The American Cinema and WWI, 1914–1941* (Rutherford, NJ, 1981); and Craig W. Campbell, *Reel America and World War One* (Jefferson, NC, 1985).

74. Although Wilson condemned the excesses and anti-German hysteria of the period he never took any strong action against them, and as the war went on the CPI's productions became increasingly virulent. On anti-German hysteria and the Wilson administration's contribution to it, see David M. Kennedy, *Over Here: The First World War and American Society* (New York, 1980).

75. *World Outlook* (November 1918), p. 8.

76. George Creel, *Complete Report of the Chairman of the Committee on Public Information* (Washington, D.C., 1920), p. 47.

77. George Creel, *Rebel At Large: Recollections of Fifty Crowded Years* (New York, 1947), p. 138; *Washington Herald*, July 19, 1918; *Washington Times*, July 19, 1918; clippings in Papers of George Creel, Manuscript Division, Library of Congress, Washington, D.C. (hereafter CP), Box 5.

78. See Creel to Charles Hart, June 6, 1918, in record group RG 63, Records of the Committee on Public Information, 1917–1919, CPI 1-A1, National Archives, Washington, D.C.; Creel to North Carolina Press Association, July 25, 1918, in CP, Bx 5. See also Creel to Wilson, undated memorandum asking Wilson to allay theater managers' fear of competition from four-minute men, CP, Bx 2. On the four-minute men, see James Mock and Cedric Larson, *Words That Won the War* (Princeton, NJ, 1939), pp. 113–126; Ward, *Motion Picture Goes to War*, p. 129.

79. *New York Times*, July 29, 1917, p. 8.

80. Creel, *How We Advertised America*, p. 5.

81. Charles S. Hart to Bruce Barton, Mar. 28, 1918, CPI 10A-A1.

82. Mould, *American Newsfilm*, pp. 268–272.

83. Howard Herrick to George Newland, April 19, 1918, CPI 10A-A1.

84. Creel, speech at premier showing of *America's Answer*, July 29, 1916, CPI, Bx 5.

85. *Variety*, Aug. 2, 1918, p. 38, quoted in Mould, *American Newsfilm*, p. 266.

86. *Fortnightly Review*, 113 (May 1920), p. 718.

87. Terry Ramsaye, *A Million and One Nights*, p. 784. On the Signal Service photographers, see Mould, *American Newsfilm*.

88. Charles S. Hart to James Hoff, April 5, 1918, CPI 10A-A1.

89. Gilbreth to O. O. Ellis, Jan. 14, 1918, pp. 4–5, GP, Reel 4.

90. Gilbreth to Ellis, Feb. 12, 1918, GP, Reel 4.

91. Frank Gilbreth to Lillian Gilbreth, Jan. 15, 18, 1918, GP, Reel 3.

92. O. O. Ellis to Margaret Hawley, Dec. 5, 1928, quoted in Margaret Hawley, "The Life of Frank Gilbreth" (M.A. thesis, University of California, Berkeley), p. 92.

93. Following Gilbreth's stay at Fort Sill, the Army hired animation expert J. R. Bray to produce a second series of training films. Bray combined actual footage with animated cross sections and slow-motion close-ups that showed the principles of the object under study. *Literary Digest*, 60 (Feb. 22, 1919), pp. 25–26.

94. *The Bookman*, 47 (May 1918), p. 238.

95. *Current Opinion*, 64 (January 1918), p. 404.

96. *The Bookman*, 47 (May 1918), p. 239.

97. *Current Opinion*, 64 (January 1918), p. 404.

VI

The Golf Stick and the Hoe

1. Thorstein Veblen, *The Instinct of Workmanship* (New York, 1914), pp. 311–319.

2. Noah Webster, *An American Dictionary of the English Language* (1828); reprint ed. (New York, 1970), vol. II; William Cragie and James Hulbert, eds., *A Dictionary of American English* (Chicago, 1946), vol. IV; *Oxford English Dictionary* (Oxford, 1934), vol. XII.

3. On the transformation of attitudes about leisure time and its use, see Lewis A. Erenberg, *Steppin' Out: New York Nightlife and the Transformation of American Culture, 1890–1930* (Chicago, 1981); Lawrence W. Levine, *Highbrow/Lowbrow. The Emergence of Cultural Hierarchy in America* (Cambridge, MA, 1988); John Kasson, *Amusing the Millions: Coney Island at the Turn of the Century* (New York, 1978); Lary May, *Screening Out the Past: The Birth of Mass Culture and the Motion Picture Industry* (Chicago, 1983); Kathy Piess, *Cheap Amusements: Working Women and Leisure in Turn-of-the-Century New York* (Philadelphia, 1986); Daniel Rodgers, *The Work Ethic in Industrial America, 1850–1920* (Chicago, 1974); and Roy Rosenzweig, *Eight Hours for What We Will: Workers and Leisure in an Industrial City, 1870–1920* (Cambridge, MA, 1985).

4. *CR*, Aug. 19, 1919, p. 3994.

5. *Cleveland Plain Dealer*, April 16, 1914, p. 13. Georgia and Florida were also particularly dissatisfied with standard time.

6. *Literary Digest,* 31 (Dec. 15, 1900), p. 725.

7. *American City,* 15 (September 1916), p. 277; *American Review of Reviews,* 54 (August 1916), pp. 208–209.

8. Franklin, "An Economical Project" (1784), in Albert Henry Smyth, ed., *The Writings of Benjamin Franklin* (New York, 1970), vol. IX, pp. 183–189. See also *The Autobiography of Benjamin Franklin,* in Smyth, ed., *Writings of Benjamin Franklin,* vol. I, p. 385. Decadent Parisians, Franklin noticed, rose late and conducted their affairs by candlelight. In an anonymous letter to the *Journal of Paris,* Franklin feigned astonishment, on "accidentally" awakening before noon one day, at the sunlight streaming through his hotel window. Awaken all of Paris at sunrise with ringing bells or booming cannons, he conjectured, and her citizens would soon be doing business by daylight, saving 96 million livres in the cost of candles. Franklin proposed waking people at sunrise, not at any specific hour of the clock. An interesting precursor to daylight saving appeared in the *Connecticut Journal,* Feb. 20, 1827, p. 2. The writer noted that some New Haven churches rang their morning bells an hour earlier in the summer, and suggested that if this custom were adopted by the city as a whole, "many thousand dollars" would be saved, in addition to "increased health and vigor to the constitution."

9. Willett's pamphlet is reprinted in Donald de Carle, *British Time* (London, 1947), pp. 151–160; and discussed in *The Washington Post,* Mar. 28, 1909. On time in Cleveland and Detroit, see *American Review of Reviews,* 54 (August 1916), pp. 206–208; Cleveland Chamber of Commerce, *Annual Reports,* 1914, 1915, 1916, pp. 145–148, 108–109, 85–86, respectively; *Cleveland Plain Dealer,* April 16, 17, 1914, p. 8; April 30, p. 1; May 2, p. 1; City of Detroit, *Journal of the Common Council,* April 27, 1915, pp. 714–715; *Detroit Free Press,* May 13, 1915, pp. 4, 13; May 16, 1915, pp. 1, 4.

10. *Cleveland Plain Dealer,* May 1, 1914, pp. 1, 7.

11. *Cleveland Plain Dealer,* May 2, 1914, p. 1; April 16, 1914, p. 13; Cleveland Chamber of Commerce, *Annual Reports,* 1914, p. 147.

12. United States, Senate, Committee on Interstate Commerce, *Standardization of Time,* May 26, 1916 (Washington, D.C., 1916).

13. *Engineering Record*, 73 (June 10, 1916), p. 759; *Literary Digest*, 52 (June 3, 1916), p. 1644; *Literary Digest*, 52 (Aug. 5, 1916), p. 300; *Saturday Evening Post*, July 29, 1916, p. 22.

14. *The Independent*, 86 (June 12, 1916), p. 430; *Engineering Record*, 73 (June 10, 1916), p. 759.

15. *New York Times*, May 8, 1916, pp. 6, 10.

16. *The Independent*, 86 (June 12, 1916), p. 430. For the most extreme example of daylight saving as a curb on nightlife, see George Patrick, "The Psychology of Daylight Saving," in *The Scientific Monthly* (November 1919), pp. 385–396.

17. *Scientific American*, 114 (June 10, 1916), p. 614; 115 (Aug. 12, 1916).

18. *New York Times*, May 8, 1916, p. 6.

19. *New York Times*, May 4, 1916, p. 16; May 6, 1916, p. 10.

20. *New York Times*, May 30, 1916, p. 5; Jan. 31, 1917, p. 9.

21. Several stacks of these cards, held together by cloth tapes, remain in the records of the Committee on Interstate Commerce in the National Archives. For documents in the Archives relating to daylight saving in this period, see H.R. 64A-H11.7 through H.R. 66A-H10.2.

22. Boston, Chamber of Commerce, Special Committee on Daylight Saving Plan, "An Hour of Light for an Hour of Night." The pamphlet is undated, but must have been issued sometime in late 1916. It was reprinted as a supplement to *Current Affairs*, March 19, 1917.

23. United States, Senate, Committee on Interstate Commerce, *Daylight Saving and Standard Time for the United States*, May 3, 10, 1917 (Washington, D.C., 1917), pp. 4–8, 26–27.

24. Senate, Committee on Interstate Commerce, *Calendar No. 53, Report No. 46: Daylight Saving* (May 25, 1917). The Committee concluded that by the time the bill passed in 1917, its period of greatest usefulness would be largely over.

25. *New York Times*, Jan. 14, 1917, p. 7; Jan. 16, 1917, p. 7; May 18, 1917, p. 7; June 5, 1917, p. 15; *The Automobile*, Feb. 8, 1917, p. 336; July 5, 1917, p. 33; *The Outlook*, 115 (Feb. 19, 1917), p. 293.

26. *New York Times*, Jan. 23, 1917, p. 11; Resolution of Pennsylvania Senate, May 16, 1917. In NA, H.R. 65A-H6.4.

27. Few examples of these posters remain. Until 1976, the New Jersey Historical Society had copies of at least five, but these were sold at auction in that year. The two posters cited were reproduced in *American City,* 18 (March 1918), p. 218, and James R. Beniger, *The Control Revolution: Technological and Economic Origins of the Information Society* (Cambridge, MA, 1986), p. 328 (thanks to Steven Reber for the latter). See also New Jersey Historical Society, *Collection of World War I Posters* (Newark, NJ, 1986), p. 30. Each of these posters bears the legend "National Cigar Stores Company." Since the themes and style of the posters are nearly identical to the National Daylight Saving Association's mailing cards, I am assuming that this company sponsored the posters in cooperation with Marks's organization. But I have no idea how many were produced or how widely they were disseminated, or if other companies sponsored similar posters.

28. For editorials favorable to daylight saving, see the *Survey,* 39 (Jan. 12, 1918), p. 420; *Scientific American,* 118 (Jan. 26, 1918), p. 82; *American City,* 18 (March 1918), pp. 217–218; *The Outlook,* 118 (Mar. 13, 1918), p. 396; *Literary Digest,* 56 (Mar. 30, 1918), p. 20.

29. Joshua Burley and Co. to H. H. Dale of New York, Mar. 11, 1918; same letter dated Mar. 12, 1918, from Schwartz and Jaffee Boys and Juvenile Clothing, both in NA, H.R. 65A-H6.4. The same letter was also entered into the *Congressional Record* by Rep. Moore of Pennsylvania on Mar. 15, 1918 (p. 3575). I do not know who originated this letter, but suspect it to be either Marks's organization or the United States Chamber of Commerce under A. Lincoln Filene. These two organizations were the most active on the bill's behalf.

30. *CR,* 56:4 (Mar. 15, 1918), pp. 3564–3584; Hoover's endorsement, p. 3579.

31. *CR,* p. 3571. Although at least two other congressmen mentioned this article, I have been unable to find a copy of it. It certainly seems too ridiculous to have been a real proposal.

32. *Appendix to the Congressional Record,* 56:4 (Mar. 18, 1918), p. 225. Hereafter *ACR.*

33. *CR,* 56:4 (Mar. 16, 1918), p. 3595. *New York Times,* Mar. 29, 1918, p. 8; Mar. 31, 1918, sec. iv, p. 14; Apr. 1, 1918, p. 11; May 9, 1918, p. 12.

34. *Current Opinion*, 64 (May 1918), p. 366; *Survey*, 41 (Oct. 26, 1918), p. 105. The bill officially expired on October 27, but it proved so popular in some circles that Congress considered extending it to apply all year long. Even its most vocal partisans pointed out that keeping daylight saving in winter made no sense, and in fact only increased the number of morning hours spent in darkness. The bill was allowed to expire on October 27, which of course made it necessary to pass an entirely new bill next year.

35. *Literary Digest*, 59 (Nov. 9, 1918), p. 12.

36. *Survey*, 41 (Nov. 9, 1918), p. 171.

37. Interstate Commerce Commission, *No. 10122. Standard Time Zone Investigation* (Washington, D.C., Oct. 1–Oct. 24, 1918), pp. 278–279, 281.

38. *Literary Digest*, 60 (Feb. 1, 1919), pp. 23–25.

39. Petition from Citizens of New Britain, Connecticut, to Senator MacLean, Feb. 27, 1919. See also Minnesota State Vegetable Growers to Schall of Minnesota, Feb. 14, 1919; Diamond State Grange of Stanton, Delaware, to Albert Polk, Feb. 10, 1919, all in NA, H.R. 65A-H6.4. This latter file contains the fullest account of opposition to daylight saving. Until recently, most of the letters from this time related to daylight saving were being kept at the Department of Transportation. When I called that department in fall of 1986, however, I was told the letters had been destroyed the year before.

40. *New York Times*, Feb. 22, 1919, p. 8; Feb. 23, 1919, p. 8; *Scientific American*, 120 (Mar. 8, 1919), p. 220; *New York Times*, Feb. 27, 1919, p. 17.

41. An exhaustive account of the arguments against daylight saving appears in House of Representatives, Committee on Interstate and Foreign Commerce, *Repeal of Section Three of the Daylight Saving Act* (Washington, D.C., June 2–3, 1919), pts. I–III; and in the *Congressional Record* and *Appendix* throughout the summer months, especially June 18, 1919, pp. 1304–1335. Congressmen reprinted a considerable number of letters and petitions from their constituents in *CR*, and many examples of these survive in the National Archives, especially record group H.R. 66A-H10.1.

42. *CR*, 58:2 (June 18, 1919), p. 1320.

43. *ACR*, 58 (June 18, 1919), p. 8885. See also *CR*, 58:2, pp. 1322–1323.

44. *ACR*, 58 (June 18, 1919), p. 8887. For an early example of these objections, see Petition from the Farmer's Club of Enfield, Connecticut, June 20, 1917, in NA record group H.R. 65A–H6.4.

45. *CR*, 58:2, p. 1305.

46. *CR*, 58:2, p. 1305; 58:3, p. 2843, 3505; and *Literary Digest*, June 14, 1919, pp. 16–17.

47. *CR*, 58:2, p. 1317; 58:3 (Aug. 1, 1919), p. 3509. It also put some places closer to the local sun time than they had been before daylight saving, assuming, of course, that standard and not local time prevailed. Longitude, and a community's location within a zone, greatly affects the period of daylight. In the South, daylight saving is less drastic than in the North, where the variation in the length of the day from summer to winter is much greater. For a description of the effects of daylight saving in different regions, see Ian Bartky and Elizabeth Harrison, "Standard and Daylight-saving Time," in *Scientific American*, 240 (May 1979), pp. 46–53.

48. Extension of Remarks of Thomas Schall of Minnesota, June 24, 1919, in *ACR*, 58, p. 8891.

49. Extension of Remarks of William Boies of Iowa, June 18, 1919, in *ACR*, 58, p. 8876.

50. *CR*, 58:2, pp. 1332–1334.

51. Wilson's first veto message is reprinted in *CR*, 58:3 (July 12, 1919), p. 2492; his second veto, dated Aug. 15, 1919, is in the NA, record group H.R. 66A–H10.

52. *CR*, 58:3 (July 12, 1919), p. 2835.

53. *CR*, 58:3 (Aug. 1, 1919), p. 3507; "Extension of Remarks of William Boies of Iowa," June 18, 1919, in *ACR*, 58, p. 8876; *CR*, 58:2 (June 18, 1919), p. 1313; *CR*, 58:2 (June 18, 1919), p. 1321.

54. *ACR*, 58, p. 8875.

55. *Literary Digest*, 60 (Feb. 1, 1919), p. 24; 62 (Sept. 6, 1919), p. 18. See also *Outlook*, 122 (June 25, 1919), p. 320–321; *Scientific American*, 20 (June 21, 1919), p. 648; *American Review of Reviews*, 60 (August 1919), p. 121; *The Review*, 1 (Sept. 6, 1919), p. 357; *New*

York Times, Aug. 21, 1919, p. 10, Aug. 22, 1919, p. 10, Aug. 25, 1919, p. 10, Oct. 6, 1919, p. 16; and the numerous letters included in *CR* during the summer of 1919.

56. This fact still holds true today. Ride any city's mass transportation system at six o'clock in the morning, and the patrons are largely black and Hispanic, dressed casually or in work clothes. Two hours later, the riders are nearly all white, and wearing suits.

57. See summary of arguments in House Committee on Interstate and Foreign Commerce, *Report No. 42: Repeal of Daylight Saving Law* (June 13, 1919), pp. 2–3.

58. *New York Times,* Aug. 22, 1919, p. 10.

59. American Federation of Labor, *Proceedings of the 39th Annual Convention,* 39 (June 10, 1919), p. 185. See also *New York Times,* June 11, 1919, p. 17. For other evidence of worker opposition, see testimony in hearings on daylight saving in House of Representatives, Committee on Interstate and Foreign Commerce, *Repeal of Section Three of the Daylight Saving Act* (Washington, D.C., June 2–3, 1919), esp. pp. 9–17, 34–39, 44–45.

60. See letter from Marks in *The American City,* 21 (September 1919), p. 212; also 21 (November 1919), pp. 432–433; *American Review of Reviews,* 60 (December 1919), pp. 647–648; *Literary Digest,* 63 (Nov. 1, 1919), pp. 17–18; 61 (April 10, 1920), pp. 30–31.

61. *Massachusetts State Grange* vs. *Benton, United States Supreme Court Reports,* 272 (1926), pp. 525–529. For a summary of changes in time since 1919, see United States Department of Transportation, *Standard Time in the United States: A History of Standard and Daylight Saving Time in the United States and an Analysis of the Related Laws* (Washington, D.C., 1970).

62. *New York Times,* Mar. 19, 1918, p. 12.

63. *New York Times,* Apr. 2, 1918, p. 15.

64. See "Statement of Edward King," in House of Representatives, Committee on Interstate and Foreign Commerce, *Repeal of Section Three of the Daylight Saving Act,* pp. 40–42, and *CR,* 52:8, pp. 1332–1334.

65. *New York Times,* Aug. 9, 1919, p. 7.

66. Scrap of film in my possession, undated but approximately 1915–19. The scrap, on highly unstable nitrate film stock, is all that remains of a longer film, which would have been used to introduce a feature, or as filler between one-reelers. Thanks to Andrea Kalas of the UCLA Film Archives, Los Angeles, for finding and sending me this fragment, and to Helena Wright of the Division of Graphic Arts at the National Museum of American History, Washington, D.C., for confirming the dating.

67. Harold B. Franklin, President, Fox West Coast Theaters, to E. B. Duerr, Pathé Studios, May 17, May 26, June 20, 1930; Franklin to George Young of the *Los Angeles Examiner*, May 23, 1930, all in Margaret Herrick Library, Academy of Motion Picture Arts and Sciences, Beverly Hills, CA, Pathé Exchange Collection, folder 11, "General Matters No. 2." See also editorial in the *Los Angeles Examiner*, May 23, 1930.

68. Fern S. Chapman, "Business's Push for More Daylight Time," in *Fortune*, 110 (Nov. 12, 1984), p. 149.

69. Telephone conversation between the author and James Benfield, Bracy, Williams and Co., 1000 Connecticut Ave. NW, Suite 304, Washington, D.C., 20036, November 1986.

70. *The Independent*, July 12, 1919, p. 51.

71. *New York Times*, Apr. 2, 1919, p. 10.

72. Orville H. Platt, "Invention and Advancement" (1891), reprinted in Henry Nash Smith, *Popular Culture and Industrialism, 1865–1890* (New York, 1967), pp. 40–41.

73. *Scientific Monthly* (November 1919), pp. 385–396.

74. Responses to relativity described in Lawrence W. Levine, "The Unpredictable Past: Reflections of Recent American Historiography," in *American Historical Review*, 94 (June 1989), p. 677.

75. Robert and Helen Merrell Lynd, *Middletown: A Study in Modern American Culture* (New York, 1929), pp. 253–254, 258.

76. *The American Mercury*, VI (October 1925), p. 160. In *Anti-Intellectualism in American Life*, *The Paranoid Style in American Politics*, and *The Age of Reform*, Hofstadter linked fundamentalism to what he saw as a long tradition of delusional thinking on the American political margins. His view deprives fundamentalists of their historical

uniqueness and, worse, characterizes their objection to modernism as pathological and paranoid largely because it failed. See also Peter Novick, *That Noble Dream: The "Objectivity Question" and the American Historical Profession* (Cambridge, MA, 1988), pp. 337–341.

77. Ernest R. Sandeen, *The Roots of Fundamentalism: British and American Millenarianism, 1800–1930* (Chicago, 1970). My understanding of fundamentalism has been greatly influenced by George M. Marsden, *Fundamentalism and American Culture: The Shaping of Twentieth Century Evangelism, 1870–1925* (New York, 1980). Marsden points out that Sandeen's emphasis on Millenarian thinking fails to account for the larger fundamentalist movement.

78. Samuel Harris quoted in Jon H. Roberts, *Darwinism and the Divine in America: Protestant Intellectuals and Organic Evolution, 1859–1900* (Madison, WI, 1988), p. 9.

79. For American responses to Darwin, see James R. Moore, *The Post-Darwinian Controversies: A Study of the Protestant Struggle to Come to Grips with Darwin in England and America, 1870–1900* (Cambridge, 1979); Roberts, *Darwinism and the Divine.*

80. Quoted in Ray Ginger, *Six Days or Forever? Tennessee v. John Tyler Scopes* (1958; reprint ed. Chicago, 1969), p. 15. The quotation reflects common-sense realism's approach to science, in that it relates religion to the natural world—observe the natural world "scientifically" for examples of the proper use of time.

81. On the history of the week, see Eviatar Zerubavel, *The Seven-Day Circle* (New York, 1985).

82. For geology's impact on ideas about time, see Stephen Jay Gould, *Time's Arrow, Time's Cycle: Myth and Metaphor in the Discovery of Geological Time* (New York, 1986), and for examples of the problems geology posed, see Henry Adams's chapter on Darwinism in *The Education of Henry Adams.* The notion that instead of "progress" evolution only gave evidence of change greatly upset Adams. He marveled at the discovery of two prehistoric fish, *terabratula* and *tserapsis,* which appeared to have survived unchanged from antiquity. If there was no progress, thought Adams, then history and time were ultimately just meaningless records of change.

83. Richard Hofstadter, *Social Darwinism in American Thought* (1944; reprint ed. Boston, 1955), p. 58. For examples of social Darwinist

thinking, see Stephen Jay Gould, *The Mismeasure of Man* (New York, 1981). For the connection between evolution, machines, and modernist culture, see Cecelia Tichi, *Shifting Gears: Technology, Literature, Culture in Modernist America* (New York, 1987), p. 34.

84. Norman F. Furniss, *The Fundamentalist Controversy, 1918–1931* (New Haven, 1954), pp. 23–25.

85. Quoted in Lawrence W. Levine, *Defender of the Faith: William Jennings Bryan: The Last Decade* (New York, 1965), p. 280.

86. On Darrow, see Ginger's uncritical *Six Days or Forever?* pp. 47–68.

87. Levine, *Defender of the Faith*, p. 339.

88. This is pure speculation on my part. According to Leslie H. Allen, *Bryan and Darrow at Dayton* (1925; reprint ed. New York, 1967), p. 157, Raulston's watch was fifteen minutes fast on the last day of the trial. Fifteen minutes error was probably too great to have been accidental. Dayton was then located in the far east of the central zone, and central standard time would have been about fifteen minutes slower than local time by the Dayton sun. If Raulston set his watch by a sundial, or by a local clock set to mean local time, he would have been about fifteen minutes fast of standard time.

89. Quoted in Ruth Miller Elson, *Guardians of Tradition: American Schoolbooks of the Nineteenth Century* (Lincoln, NE, 1964), p. 16.

90. This is according to Ray Ginger, *Six Days or Forever?* p. 170. The nearly complete transcript of the Bryan-Darrow confrontation given in Leslie H. Allen, *Bryan and Darrow at Dayton*, does not record any mention of the specific time of creation, but does record it given as part of the scientific testimony read earlier in the trial. The comment about "eastern standard time" may have occurred then.

91. Audience reaction quoted in L. Sprague DeCamp, *The Great Monkey Trial* (New York, 1968), p. 403.

92. The 1960 movie, directed by Stanley Kramer, is oddly obsessed with time as well. It opens with a shot of the clock over the town courthouse approaching 8:00 a.m., then moves to a shot of three local citizens, all looking at their watches. Across the square they meet a fourth, also looking at his watch. In the trial scenes the judge sits before an open window, through which the camera shows a building with the painted sign "Traveler's Time." At one point in his cross-

examination of the Bryan character, Spencer Tracy refers to foes of evolution as "clock-stoppers." Kramer seemed to pick up the theme of natural time versus industrial, clock-based, standardized time.

93. Marsden, *Fundamentalism and American Culture,* p. 214.

94. Quoted in Leslie H. Allen, *Bryan and Darrow at Dayton,* p. 190.

95. Ibid., p. 195.

Epilogue

1. Standard time was first made federal law in 1918. The Uniform Time Act of 1966 extended the federal government's authority over time and established penalties for failure to conform to standard time.

2. For a discussion of the relation between the internal, biological clock and social structures, see Michael Young, *The Metronomic Society: Natural Rhythms and Human Timetables* (Cambridge, MA, 1988).

Index

Page numbers in *italics* refer to illustrations.